Targeting Mental Maths

AUSTRALIAN CURRICULUM EDITION

Garda Turner

AF583957

PASCAL PRESS

Copying for educational purposes
The Australian *Copyright Act 1968* (the Act) allows a maximum of one chapter or 10% of this book, whichever is greater, to be copied by any educational institution for its educational purposes provided that the educational institution (or the body that administers it) has given a remuneration notice to Copyright Agency under the Act.

For details of the Copyright Agency licence for educational institutions contact:

Copyright Agency
Level 12, 66 Goulburn Street
Sydney NSW 2000
Telephone: (02) 9394 7600
Facsimile: (02) 9394 7601
E-mail: memberservices@copyright.com.au

Copying for other purposes
Except as permitted under the Act (for example, any fair dealing for the purpose of study, research, criticism or review) no part of this book may be reproduced, stored in a retrieval system, or transmitted in any for or by any means without prior written permission. All inquiries should be made to the publisher.

Targeting Mental Maths Year 5
Written by Garda Turner
Copyright © Blake Publishing 2012
Updated in 2013 for the Australian Curriculum
Reprinted 2014, 2015, 2016, 2017, 2018, 2020, 2021
Updated in 2023 for the Australian Curriculum
Reprinted 2024

ISBN 978-1-92288-727-6

Published by Pascal Press
PO Box 250
Glebe NSW 2037
(02) 9198 1748
www.pascalpress.com.au

Publisher: Katy Pike
Series Editor: Garda Turner
Editor: Amanda Santamaria
Printed by Wai Man Book Binding (China) Ltd.

Contents

Introduction

The importance of mental mathematics

The development of a variety of mental strategies helps to make children confident mathematicians.

Students who can perform mathematical computations swiftly and accurately in their heads are seen as being 'good at maths'. In general, poorer performing students use less efficient strategies and the development of mental computation through a strategy approach can help them move forwards. Research has shown that 'after instruction students seemed more likely to use strategies that reflected number sense and that this was a long-term change.' (Markovits and Sowder, 1994)

The Year 5 Targeting Mental Maths book has been written to complement the Targeting Maths Australian Curriculum Year 5 Student Book. The two-page weekly units run parallel to the contents in the student book. Units are divided into four terms of work and each term ends with a Revision Unit. There are 35 units in total.

A unit consists of two facing pages. The left-hand page has three groups of quick mental exercises that constantly practise the necessary maths facts. The right-hand page always starts with an explanation and practice of a mental strategy. This is followed by exercises that include work on space, measurement, position, number, data, chance etc. The page concludes with a Problem of the week.

Introduction

Students with number sense know the relative size of numbers and use a variety of computation strategies to solve a problem. This enables them to approach maths with a real sense of what they are being asked to do, making them more confident, with a greater ability to work mathematically.

Targeting Mental Maths fulfils the need for easily accessible mental warm-ups, constant practice of maths facts, a useful homework book and further revision of a concept being taught.

Sets A, B and C
The left-hand page has three groups of quick mental exercises that constantly practise the necessary maths facts.

Two-page units
A unit consists of two facing pages.

Mental Strategies
The right-hand page always starts with an explanation and practice of a mental strategy. This structured focus of mental strategies is unique to the Targeting Mental Maths series.

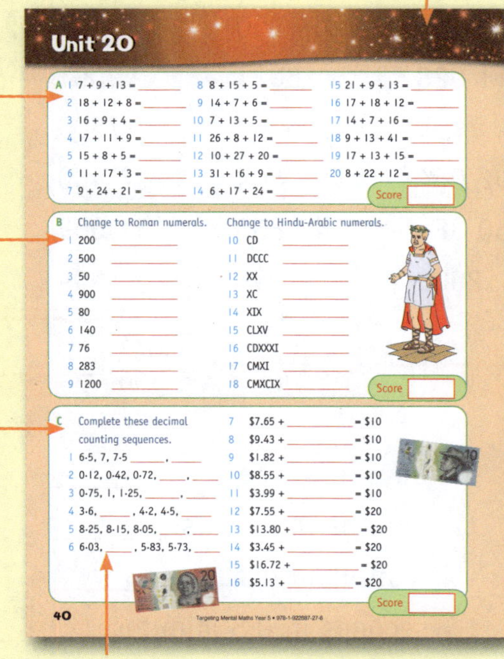

Unit 20

A
1 7 + 9 + 13 = ____ 8 8 + 15 + 5 = ____ 15 21 + 9 + 13 = ____
2 18 + 12 + 8 = ____ 9 14 + 7 + 6 = ____ 16 17 + 18 + 12 = ____
3 16 + 9 + 4 = ____ 10 7 + 13 + 5 = ____ 17 14 + 7 + 16 = ____
4 17 + 11 + 9 = ____ 11 26 + 8 + 12 = ____ 18 9 + 13 + 41 = ____
5 15 + 8 + 5 = ____ 12 10 + 27 + 20 = ____ 19 17 + 13 + 15 = ____
6 11 + 17 + 3 = ____ 13 31 + 16 + 9 = ____ 20 8 + 22 + 12 = ____
7 9 + 24 + 21 = ____ 14 6 + 17 + 24 = ____
Score

B Change to Roman numerals. Change to Hindu-Arabic numerals.
1 200 ____ 10 CD ____
2 500 ____ 11 DCCC ____
3 50 ____ 12 XX ____
4 900 ____ 13 XC ____
5 80 ____ 14 XIX ____
6 140 ____ 15 CLXV ____
7 76 ____ 16 CDXXXI ____
8 283 ____ 17 CMXI ____
9 1200 ____ 18 CMXCIX ____
Score

C Complete these decimal counting sequences.
1 6·5, 7, 7·5 ____, ____
2 0·12, 0·42, 0·72, ____, ____
3 0·75, 1, 1·25, ____, ____
4 3·6, ____, 4·2, 4·5, ____
5 8·25, 8·15, 8·05, ____, ____
6 6·03, ____, 5·83, 5·73, ____
7 $7.65 + ____ = $10
8 $9.43 + ____ = $10
9 $1.82 + ____ = $10
10 $8.55 + ____ = $10
11 $3.99 + ____ = $10
12 $7.55 + ____ = $20
13 $13.80 + ____ = $20
14 $3.45 + ____ = $20
15 $16.72 + ____ = $20
16 $5.13 + ____ = $20
Score

40 Targeting Mental Maths Year 5 • 978-1-922887-27-6

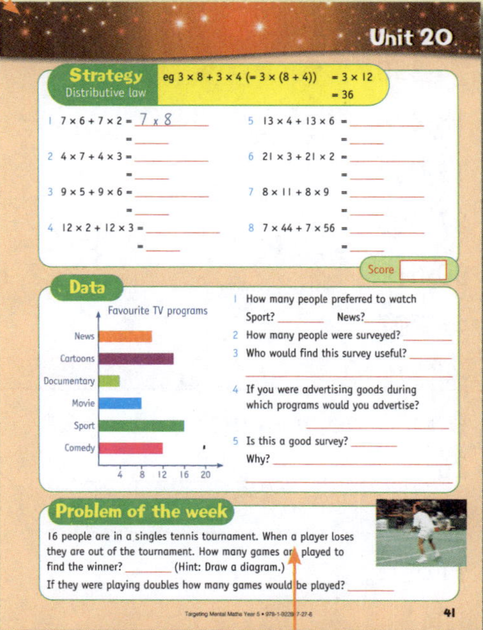

Unit 20

Strategy
Distributive law
eg 3 × 8 + 3 × 4 (= 3 × (8 + 4)) = 3 × 12 = 36

1 7 × 6 + 7 × 2 = 7 x 8 = ____
2 4 × 7 + 4 × 3 = ____ = ____
3 9 × 5 + 9 × 6 = ____ = ____
4 12 × 2 + 12 × 3 = ____ = ____
5 13 × 4 + 13 × 6 = ____ = ____
6 21 × 3 + 21 × 2 = ____ = ____
7 8 × 11 + 8 × 9 = ____ = ____
8 7 × 44 + 7 × 56 = ____ = ____
Score

Data

1 How many people preferred to watch Sport? ____ News? ____
2 How many people were surveyed? ____
3 Who would find this survey useful? ____
4 If you were advertising goods during which programs would you advertise? ____
5 Is this a good survey? ____ Why? ____

Problem of the week
16 people are in a singles tennis tournament. When a player loses they are out of the tournament. How many games are played to find the winner? ____ (Hint: Draw a diagram.)
If they were playing doubles how many games would be played? ____

Targeting Mental Maths Year 5 • 978-1-922887-27-6 41

Sets of questions focusing on particular sub-skills help students to learn and retain essential basic facts.

Each set is clear and uncluttered.

Covers all strands
This section includes revision of space, measurement, position, number, chance and data concepts.

Problem of the week
An interesting problem to get students thinking and applying their mathematical skills.

Answers

Answer pages are in the centre of the book so that the complete section can be easily pulled out and kept separate if the teacher so wishes.

Unit 1

A

1. 5 × 0 = ______
2. 7 × 3 = ______
3. 9 × 7 = ______
4. 5 × 5 = ______
5. 9 × 9 = ______
6. 8 × 8 = ______
7. 6 × 4 = ______
8. 7 × 7 = ______
9. 8 × 6 = ______
10. 5 × 8 = ______
11. 8 × 9 = ______
12. 7 × 8 = ______
13. 9 × 6 = ______
14. 6 × 6 = ______
15. 6 × 7 = ______
16. 9 × 8 = ______
17. 7 × 9 = ______
18. 5 × 9 = ______
19. 8 × 4 = ______
20. 6 × 8 = ______

Score

B

Value of the underlined digit.

1. 53·$\underline{6}$84 ______
2. 39·05$\underline{6}$ ______
3. 8$\underline{2}$·501 ______
4. $\underline{7}$4·390 ______
5. 18·2$\underline{7}$8 ______
6. 13·0$\underline{6}$5 ______
7. $\underline{2}$7·483 ______
8. 90·$\underline{6}$19 ______
9. 45·70$\underline{2}$ ______

Circle the largest number.

10	8994	80·002	18 058
11	77·305	97·35	75 370
12	0·5091	0·1950	9015
13	11 028	8021	2081
14	7005	5007	7500
15	92 064	92 604	92 460
16	50 386	49 860	51 009
17	2003	1998	1909

Score

C

What type of angle?

1. 26° ______
2. 90° ______
3. 112° ______
4. 104° ______
5. 8° ______
6. 91° ______
7. 179° ______
8. 89° ______
9. 100° ______
10. 14° ______
11. days in 5 weeks ______
12. months in 5 years ______
13. months in autumn ______
14. seconds in 0·5 minutes ______
15. days in a leap year ______
16. minutes in $3\frac{1}{2}$ hours ______
17. days in July ______
18. hours in 4 days ______
19. years in a decade ______
20. years in a millenium ______

Score

Targeting Mental Maths Year 5 • 978-1-922887-27-6

Strategy
Near doubles

eg 12 + 14 = (12 + 12) + 2
= 26
18 + 17 = (18 + 18) – 1
= 35

1 15 + 16 = ______
= ______

2 20 + 19 = ______
= ______

3 14 + 13 = ______
= ______

4 19 + 18 = ______
= ______

5 25 + 26 = ______
= ______

6 32 + 33 = ______
= ______

7 36 + 35 = ______
= ______

8 27 + 26 = ______
= ______

9 34 + 33 = ______
= ______

10 45 + 46 = ______
= ______

Score

Number

Write the number.

1
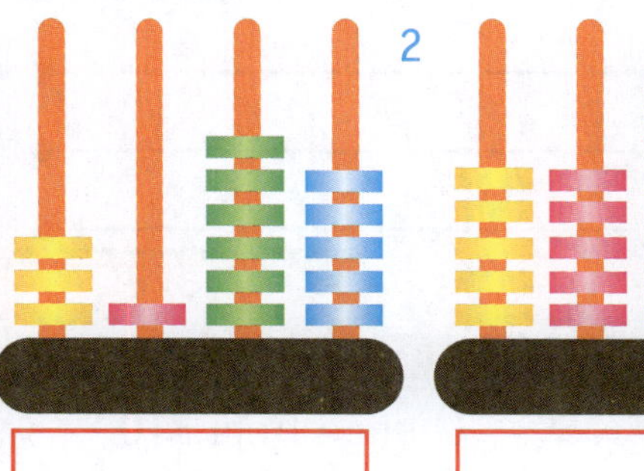

2

Draw the number.

3
2074

4

5608

Angles

Colour the right angles red, the acute angles blue and the obtuse angles green.

1 2 3
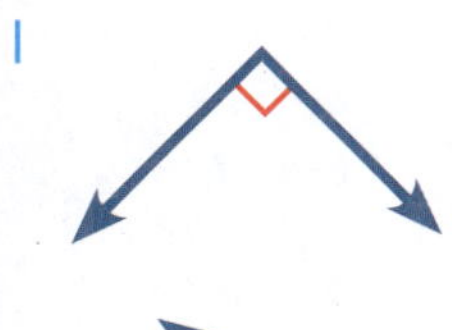
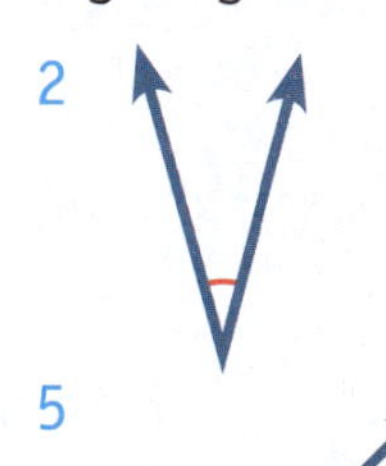

4 5 6

Problem of the week

The sum of two numbers is 162.

Their difference is 34.

What are the two numbers?

______ ______

Unit 2

A

1. 64 ÷ 8 = ________
2. 42 ÷ 7 = ________
3. 25 ÷ 5 = ________
4. 72 ÷ 9 = ________
5. 60 ÷ 6 = ________
6. 36 ÷ 6 = ________
7. 63 ÷ 9 = ________
8. 45 ÷ 9 = ________
9. 54 ÷ 6 = ________
10. 10 ÷ 10 = ________
11. 48 ÷ 8 = ________
12. 24 ÷ 4 = ________
13. 81 ÷ 9 = ________
14. 49 ÷ 7 = ________
15. 28 ÷ 7 = ________
16. 54 ÷ 9 = ________
17. 18 ÷ 3 = ________
18. 32 ÷ 8 = ________
19. 100 ÷ 10 = ________
20. 1 ÷ 1 = ________

Score

B Use the number lines.

1. 209 – 156 = ________
2. 517 – 361 = ________
3. 962 – 458 = ________
4. 723 – 516 = ________
5. 301 – 178 = ________
6. 454 – 239 = ________
7. 836 – 697 = ________
8. 640 – 443 = ________

Score

C Find the perimeter.

1. square with 4 cm sides ________
2. square with 9 m sides ________
3. square with 6 km sides ________
4. square with 12 mm sides ________
5. square with 36 cm sides ________
6. rectangle with 4 m and 7 m sides ________
7. rectangle with 9 mm and 13 mm sides ________
8. rectangle with 8 km and 15 km sides ________
9. rectangle with 19 cm and 7 cm sides ________
10. rectangle with 28 m and 16 m sides ________

Change from $10.

11. $1.40 ________
12. $6.80 ________
13. $9.50 ________
14. $4.20 ________
15. $3.15 ________
16. $1.75 ________
17. $5.35 ________
18. $8.25 ________
19. $9.95 ________
20. $7.65 ________

Score

Unit 2

Strategy
Split strategy

eg 62 + 27 = 60 + 20 + 2 + 7
= 89

1 46 + 85 = ______
= ______

2 93 + 29 = ______
= ______

3 68 + 56 = ______
= ______

4 57 + 86 = ______
= ______

5 77 + 29 = ______
= ______

6 69 + 24 = ______
= ______

7 85 + 74 = ______
= ______

8 95 + 48 = ______
= ______

Score

Time
On the other clock draw or write the time 10 minutes later.

1
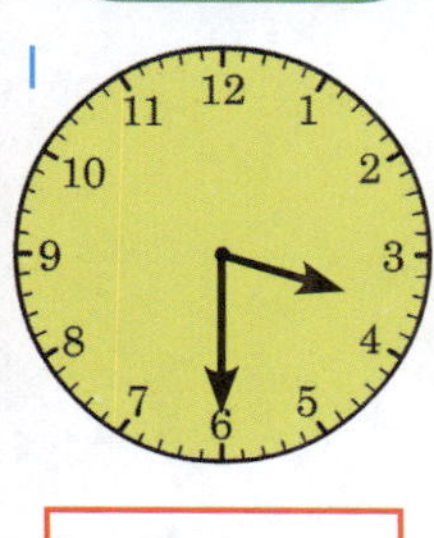
:

2
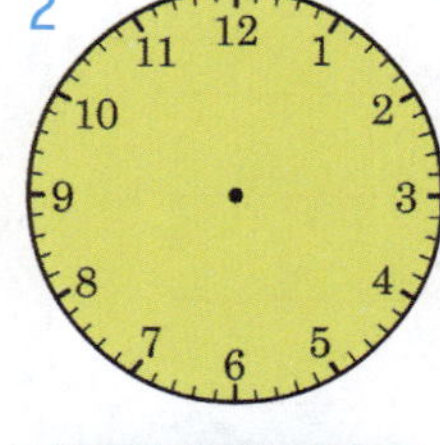
9:50

3

:

4
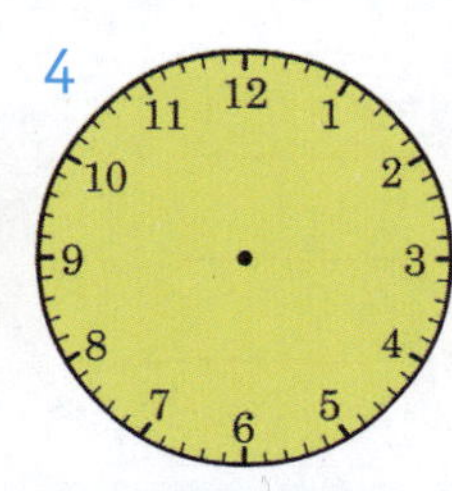
4:05

5
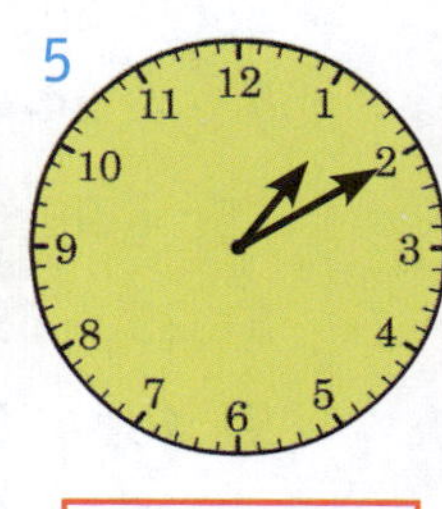
:

Chance
Write an event to describe each card on this probability line.

A B C

impossible  certain

A ______

B ______

C ______

Problem of the week
From noon to midnight how many times do the hands of the clock form a straight line? ______

Unit 3

A

1 4 + 8 = ______
2 8 + 9 = ______
3 6 + 7 = ______
4 12 + 8 = ______
5 4 + 7 = ______
6 11 + 9 = ______
7 6 + 8 = ______
8 7 + 4 = ______
9 13 + 8 = ______
10 7 + 6 = ______
11 4 + 5 = ______
12 9 + 6 = ______
13 8 + 7 = ______
14 4 + 9 = ______
15 9 + 9 = ______
16 5 + 7 = ______
17 16 + 9 = ______
18 7 + 8 = ______
19 5 + 9 = ______
20 19 + 5 = ______

Score

B Estimate only.

1 35c + 80c = ______
2 20c + 95c = ______
3 85c + 60c = ______
4 70c + 85c = ______
5 15c + 90c = ______
6 $1.25 + 55c = ______
7 $1.40 + 85c = ______
8 $1.95 + 70c = ______
9 $2.55 + 65c = ______
10 58 + 49 ______
11 37 + 81 ______
12 91 + 65 ______
13 66 + 103 ______
14 42 + 211 ______
15 115 + 172 ______
16 130 + 98 ______
17 179 + 64 ______
18 203 + 136 ______

Score

C Calculate how many:

1 0·5 of 1 year ______
2 0·5 of 1 dozen ______
3 $\frac{1}{2}$ of $2.50 ______
4 $\frac{1}{4}$ of $4.80 ______
5 0·25 of 1 hour ______
6 0·1 of 1 century ______
7 0·5 of 1 fortnight ______
8 $\frac{1}{5}$ of 5 dozen ______
9 0·25 of 1 minute ______
10 0·75 of 1 m ______
11 days in 9 weeks? ______
12 hours in 3 days? ______
13 minutes in 4 hours? ______
14 seconds in $2\frac{1}{2}$ minutes? ______
15 years in a $\frac{1}{4}$ of a millenium? ______
16 centimetres in $7\frac{1}{2}$ m? ______
17 grams in 45 kg? ______
18 millilitres in $8\frac{1}{2}$ L? ______
19 degrees in 5 right angles? ______
20 cents in $\$9\frac{1}{4}$? ______

Score

Unit 3

Strategy
Compensation strategy

eg 73 + 48 = 73 + 50 − 2
= 121
59 + 32 = 59 + 30 + 2
= 91

1 27 + 39 = ______
= ______

2 54 + 61 = ______
= ______

3 38 + 58 = ______
= ______

4 67 + 22 = ______
= ______

5 98 + 78 = ______
= ______

6 29 + 52 = ______
= ______

7 55 + 71 = ______
= ______

8 88 + 69 = ______
= ______

9 108 + 73 = ______
= ______

Score

Number

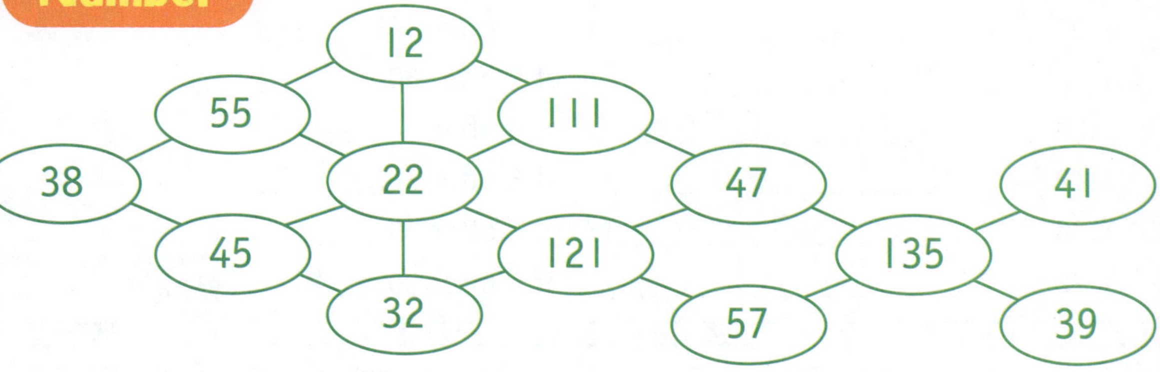

Start at 38. Colour the correct path. 38 + 17 − 33 + 99 − 74 + 88 − 96

Money

Sven bought 2 of each watch.

How much did he spend?

______

How much change did he get from $600?

______

Problem of the week

Stacey scored 79 on her first test and 93 on her second test. If she scored 250 across three tests, what did she score on the third test?

______

Unit 4

A

1 20 − 8 = ______
2 12 − 7 = ______
3 19 − 9 = ______
4 16 − 7 = ______
5 20 − 9 = ______
6 18 − 13 = ______
7 20 − 11 = ______
8 19 − 8 = ______
9 17 − 9 = ______
10 18 − 18 = ______
11 12 − 3 = ______
12 20 − 12 = ______
13 14 − 9 = ______
14 18 − 9 = ______
15 19 − 7 = ______
16 15 − 8 = ______
17 18 − 12 = ______
18 13 − 6 = ______
19 20 − 14 = ______
20 19 − 5 = ______

Score

B Write all the factors.

1 16 ______
2 24 ______
3 63 ______
4 56 ______
5 28 ______
6 42 ______
7 36 ______
8 72 ______

Write the next multiple:

9 of 6 after 20. ______
10 of 8 after 50. ______
11 of 3 after 20. ______
12 of 7 after 60. ______
13 of 9 after 50. ______
14 of 5 after 70. ______
15 of 6 after 40. ______
16 of 4 after 30. ______

Score

C Write 3 notes to make:

1 $35 ______
2 $25 ______
3 $75 ______
4 $20 ______
5 $90 ______
6 $110 ______
7 $155 ______
8 $115 ______
9 $220 ______
10 $65 ______

Time 20 minutes after:

11 $\frac{1}{2}$ past 4 ______
12 $\frac{1}{4}$ past 2 ______
13 20 past 3 ______
14 10 to 7 ______
15 $\frac{1}{4}$ to 2 ______

Time 20 minutes before:

16 $\frac{1}{4}$ past 6 ______
17 10 to 11 ______
18 25 past 9 ______
19 $\frac{1}{4}$ to 1 ______
20 5 past 3 ______

Score

Targeting Mental Maths Year 5 • 978-1-922887-27-6

Unit 4

Strategy

Jump strategy

29 + 35 = 29 + 30 + 5
= 64

61 − 36 = 61 − 30 − 6
= 25

1 48 + 37 = ______
= ______

2 95 + 66 = ______
= ______

3 88 + 54 = ______
= ______

4 53 + 77 = ______
= ______

5 66 + 47 = ______
= ______

6 81 − 24 = ______
= ______

7 92 − 65 = ______
= ______

8 64 − 37 = ______
= ______

9 73 − 28 = ______
= ______

Score

Fractions

Colour the fraction.

1 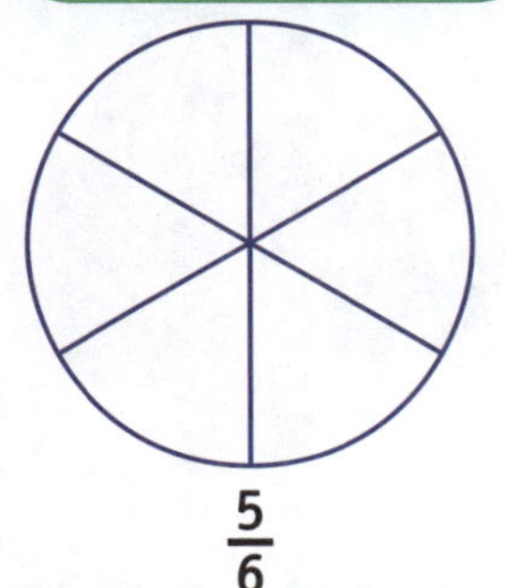

$\frac{5}{6}$

2 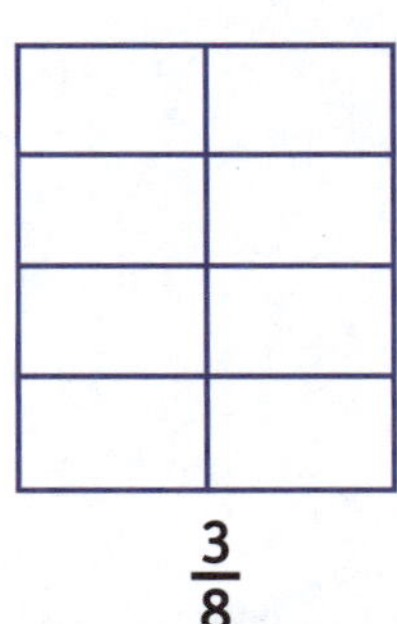

$\frac{1}{3}$

3 $\frac{3}{8}$

4 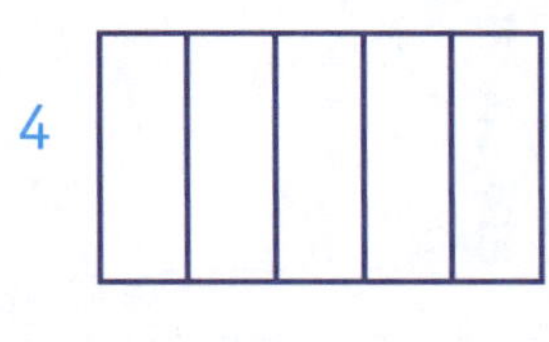 $\frac{2}{5}$

5 $\frac{7}{10}$

Space

Draw and colour a tessellating pattern.

Number

Complete the table.

Factor	6	7		8	6	
Factor	9		3		7	4
Product		56	21	64		36

Problem of the week

What number am I?

When divided by 2, 3, 4, or 5 there is nothing left over. I am the smallest possible number.

I am ______ .

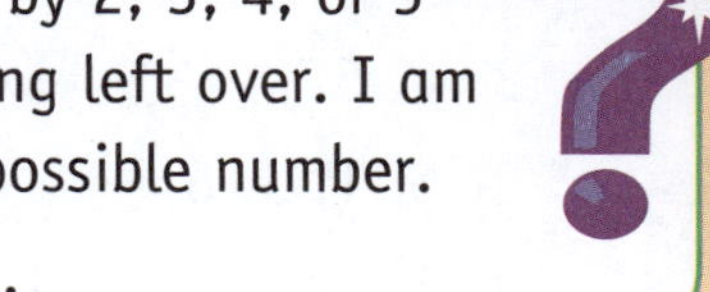

Unit 5

A

1. $7 \times 6 =$ ______
2. $35 \div 5 =$ ______
3. $7 \times 7 =$ ______
4. $24 \div 3 =$ ______
5. $8 \times 6 =$ ______
6. $7 \times 8 =$ ______
7. $63 \div 9 =$ ______
8. $8 \times 7 =$ ______
9. $80 \div 8 =$ ______
10. $28 \div 4 =$ ______
11. $7 \times 9 =$ ______
12. $9 \times 6 =$ ______
13. $72 \div 9 =$ ______
14. $6 \times 6 =$ ______
15. $9 \times 5 =$ ______
16. $6 \times 1 =$ ______
17. $0 \times 9 =$ ______
18. $8 \times 8 =$ ______
19. $36 \div 6 =$ ______
20. $9 \times 3 =$ ______

Score

B Write as a decimal.

1. $\frac{3}{10}$ ______
2. $\frac{5}{10}$ ______
3. $\frac{2}{10}$ ______
4. $\frac{7}{10}$ ______
5. $\frac{9}{10}$ ______

Write as a fraction.

6. 0·4 ______
7. 0·8 ______
8. 0·3 ______
9. 0·6 ______

Use <, > or =.

10. $\frac{1}{10}$ ______ 0·1
11. $\frac{1}{2}$ ______ $\frac{1}{10}$
12. $\frac{1}{4}$ ______ $\frac{1}{8}$
13. 0·6 ______ $\frac{6}{100}$
14. $\frac{1}{3}$ ______ $\frac{1}{6}$
15. $\frac{6}{8}$ ______ $\frac{3}{4}$
16. $\frac{2}{5}$ ______ $\frac{3}{10}$
17. $\frac{1}{2}$ ______ $\frac{4}{8}$
18. $\frac{2}{10}$ ______ $\frac{2}{5}$

Score

C Look for patterns.

1. 9, 17, 25, 33 ______
2. 1, 3, 9, 27 ______
3. 26, 22, ______, 14
4. 160, ______, 40, 20, 10
5. $\frac{1}{4}$, $\frac{1}{5}$, ______, $\frac{1}{7}$, $\frac{1}{8}$
6. 0·75, 0·7, ______, 0·6, 0·55
7. 11, ______, 29, 38, 47
8. 6, $5\frac{1}{2}$, ______, $4\frac{1}{2}$, 4
9. 0, 1, 3, 6, ______
10. 1, 4, ______, 16, 25

Colour.

11. $\frac{3}{8}$ of A red.
12. $\frac{3}{4}$ of B blue.
13. $\frac{2}{6}$ of C green.
14. $\frac{2}{3}$ of D yellow.
15. $\frac{1}{4}$ of A green.
16. $\frac{1}{3}$ of C blue.
17. $\frac{1}{8}$ of 2 dozen = ______
18. $\frac{1}{3}$ of 2 dozen = ______
19. $\frac{1}{5}$ of 40 = ______
20. $\frac{1}{6}$ of 36 = ______

A B C D

Score

Unit 5

Strategy

To multiply by 4

double, double

eg $12 \times 4 = 12 \times 2 \times 2$
$= 24 \times 2$
$= 48$

1. $16 \times 4 =$ ______
2. $13 \times 4 =$ ______
3. $15 \times 4 =$ ______
4. $19 \times 4 =$ ______
5. $22 \times 4 =$ ______
6. $50 \times 4 =$ ______
7. $110 \times 4 =$ ______
8. $61 \times 4 =$ ______
9. $32 \times 4 =$ ______
10. $75 \times 4 =$ ______

Score

Patterns

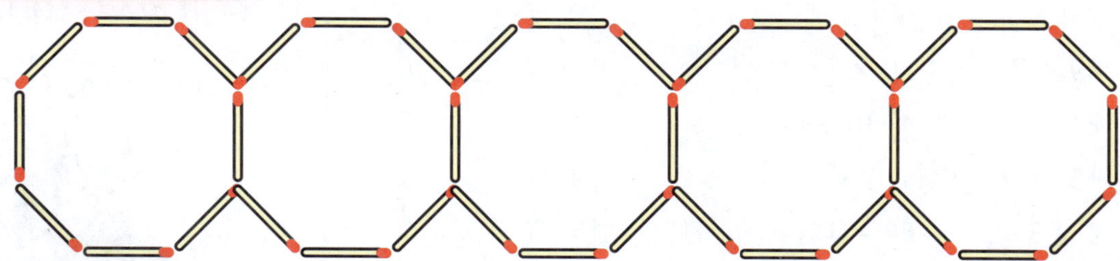

1 Fill in the table.

Number of shapes								
Number of sticks								

2 Describe the pattern in words.

______

Angles

Name each type of angle.

1
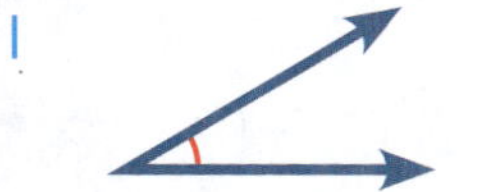

______

2
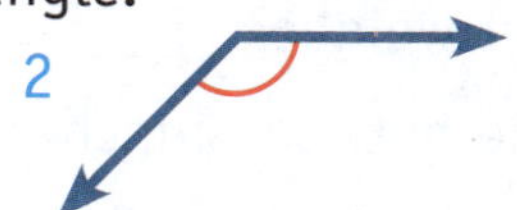

______

3
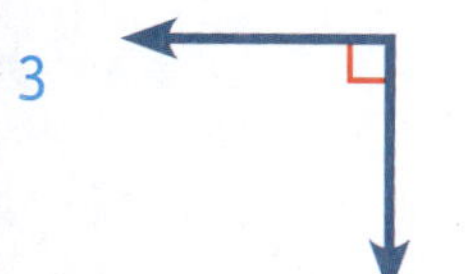

______

4
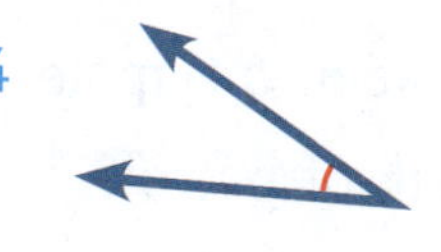

______

Problem of the week

In exactly 1080 seconds a new year will begin.

1 What is the time now? ______

2 What day is it ?

______

Unit 6

A

1 13 + 9 = ______
2 28 − 13 = ______
3 24 + 7 = ______
4 33 − 26 = ______
5 65 − 42 = ______
6 36 + 8 = ______
7 53 + 9 = ______
8 27 + 5 = ______
9 40 − 15 = ______
10 31 − 18 = ______
11 39 + 4 = ______
12 70 − 49 = ______
13 68 + 7 = ______
14 23 − 18 = ______
15 17 + 15 = ______
16 24 + 19 = ______
17 64 − 37 = ______
18 31 + 15 = ______
19 52 − 28 = ______
20 26 + 27 = ______

Score

B Write >, < or =.

1 7 + 8 ______ 5 × 3
2 49 ÷ 7 ______ 0 × 7
3 21 − 14 ______ 36 ÷ 6
4 15 + 16 ______ 4 × 8
5 9 × 8 ______ 60 + 12
6 36 − 13 ______ 14 + 11
7 5 × 8 ______ 6 × 7
8 9 × 9 ______ 100 − 9
9 25 + 14 ______ 6 × 8
10 $\frac{1}{2}$ ______ $\frac{3}{8}$
11 $\frac{1}{4}$ ______ $\frac{3}{12}$
12 $\frac{1}{5}$ ______ $\frac{2}{10}$
13 $\frac{1}{3}$ ______ $\frac{1}{4}$
14 $\frac{1}{5}$ ______ $\frac{1}{4}$
15 $\frac{3}{4}$ ______ $\frac{5}{10}$
16 $\frac{1}{3}$ ______ $\frac{1}{2}$
17 $\frac{3}{5}$ ______ $\frac{5}{10}$
18 $\frac{3}{4}$ ______ $\frac{6}{8}$

Score

C Prime number after:

1 7 ______
2 12 ______
3 24 ______
4 32 ______
5 3000 m = ______ km
6 1·5 km = ______ m
7 900 cm = ______ m
8 $4\frac{1}{2}$ m = ______ cm
9 40 mm = ______ cm
10 $3\frac{1}{2}$ cm = ______ mm

What unit of measurement is used to measure:

11 milk in a carton? ______
12 distance between two towns? ______
13 your height? ______
14 your weight? ______
15 time between lunch and dinner? ______
16 length of a pencil? ______
17 length of a garden? ______
18 weight of an apple? ______
19 a tank of petrol? ______
20 weight of an elephant? ______

Score

Targeting Mental Maths Year 5 • 978-1-922887-27-6

Strategy

To multiply by 8

double, double, double

eg $17 \times 8 = (17 \times 2) \times 2 \times 2$
$= (34 \times 2) \times 2$
$= 68 \times 2$
$= 136$

1 $16 \times 8 =$ ______
2 $13 \times 8 =$ ______
3 $32 \times 8 =$ ______
4 $18 \times 8 =$ ______
5 $25 \times 8 =$ ______
6 $31 \times 8 =$ ______
7 $19 \times 8 =$ ______
8 $42 \times 8 =$ ______
9 $15 \times 8 =$ ______
10 $51 \times 8 =$ ______

Score

Length

Measure these lines in mm.

1

2

3

4

Draw these lines accurately. 5 6 cm 6 $3\frac{1}{2}$ cm 7 56 mm 8 49 mm

Area

1 Draw a shape with an area of 15 cm^2.

2 What is the perimeter? ______

Problem of the week

A worm is at the bottom of a 6 m deep hole.

Every day it crawls up 2 m and then falls back 1 m.

How long will it take for the worm to crawl out of the hole?

______

Unit 7

A

1 7 + 6 + 7 = ________
2 9 + 3 + 8 = ________
3 4 + 5 + 11 = ________
4 3 + 7 + 10 = ________
5 1 + 6 + 13 = ________
6 5 + 4 + 9 = ________
7 8 + 11 + 6 = ________
8 7 + 12 + 9 = ________
9 8 + 3 + 12 = ________
10 4 + 7 + 11 = ________
11 5 + 8 + 9 = ________
12 9 + 4 + 7 = ________
13 6 + 7 + 8 = ________
14 3 + 6 + 12 = ________
15 7 + 8 + 9 = ________
16 8 + 8 + 7 = ________
17 5 + 9 + 6 = ________
18 9 + 9 + 7 = ________
19 6 + 8 + 11 = ________
20 7 + 4 + 13 = ________

Score

B Write the decimals.

1 15 mm = ________ cm
2 170 cm = ________ m
3 63 mm = ________ cm
4 221 cm = ________ m
5 81 mm = ________ cm
6 804 cm = ________ m
7 113 mm = ________ cm
8 35 cm = ________ m
9 3·6 cm = ________ mm
10 8·9 cm = ________ mm
11 27·4 cm = ________ mm
12 3·27 m = ________ cm
13 5·84 m = ________ cm
14 9·06 m = ________ cm
15 0·53 m = ________ cm
16 0·2 cm = ________ mm

Score

C True or false. I would use:

1 cm to measure a book. ________
2 g to weigh a dog. ________
3 L to measure petrol. ________
4 mm to measure a giant. ________
5 mL to measure medicine. ________
6 kg to weigh you. ________
7 m to measure a room. ________
8 km to measure the playground. ________
9 L to measure a cup of tea. ________
10 g to weigh a feather. ________

Acute, obtuse or right angle.

11 90° ________
12 30° ________
13 70° ________
14 101° ________
15 179° ________
16 3° ________
17 89° ________
18 19° ________
19 120° ________
20 91° ________

Score

Targeting Mental Maths Year 5 • 978-1-922887-27-6

Unit 7

Strategy
Look for patterns

eg 8 + 6 = 14
18 + 6 = 24
58 + 6 = 64

1 6 + 7 = ______
16 + 7 = ______
46 + 7 = ______

2 9 + 4 = ______
19 + 4 = ______
89 + 4 = ______

3 7 + 5 = ______
17 + 5 = ______
67 + 5 = ______

4 4 + 7 = ______
14 + 7 = ______
34 + 7 = ______

5 9 + 8 = ______
19 + 8 = ______
89 + 8 = ______

6 5 + 9 = ______
15 + 9 = ______
75 + 9 = ______

7 8 + 5 = ______
18 + 5 = ______
58 + 5 = ______

8 3 + 7 = ______
13 + 7 = ______
53 + 7 = ______

9 6 + 9 = ______
16 + 9 = ______
86 + 9 = ______

Score

Space

Draw.

1 triangular prism 2 square pyramid 3 pentagonal prism

Length

1 Write the lengths on the sides. 2 What is the perimeter? ______

Problem of the week

12 friends are having pizza for dinner.
Everyone is hungry and will eat a third of a pizza.
How many pizzas should they order? ______

Unit 8

A

1. 24 − 8 − 3 = ______
2. 19 − 7 − 8 = ______
3. 20 − 8 − 3 = ______
4. 25 − 9 − 3 = ______
5. 16 − 7 − 4 = ______
6. 18 − 3 − 5 = ______
7. 23 − 9 − 4 = ______
8. 21 − 13 − 5 = ______
9. 28 − 9 − 11 = ______
10. 17 − 6 − 5 = ______
11. 19 − 2 − 8 = ______
12. 20 − 15 − 5 = ______
13. 25 − 17 − 1 = ______
14. 27 − 6 − 18 = ______
15. 22 − 7 − 6 = ______
16. 18 − 4 − 12 = ______
17. 21 − 12 − 6 = ______
18. 24 − 8 − 7 = ______
19. 26 − 9 − 8 = ______
20. 30 − 15 − 6 = ______

Score

B

1. $\frac{1}{4} + \frac{1}{4} =$ ______
2. $\frac{1}{2} + \frac{1}{2} =$ ______
3. $\frac{1}{5} + \frac{3}{5} =$ ______
4. $\frac{5}{8} + \frac{2}{8} =$ ______

5. $\frac{3}{12} + \frac{4}{12} =$ ______

6. $\frac{3}{4} - \frac{1}{4} =$ ______

7. $\frac{10}{12} - \frac{3}{12} =$ ______
8. $\frac{2}{3} - \frac{1}{3} =$ ______

9. $\frac{4}{5} - \frac{2}{5} =$ ______

10. $\frac{1}{8}$ of 24 apples ______
11. $\frac{1}{6}$ of 60c ______
12. $\frac{1}{5}$ of $1 ______
13. $\frac{1}{4}$ of 16 books ______
14. $\frac{1}{3}$ of 36 sweets ______
15. $\frac{1}{12}$ of 1 dozen eggs ______
16. $\frac{1}{4}$ of $100 ______
17. $\frac{1}{4}$ of 28 bananas ______
18. $\frac{1}{5}$ of a century ______

Score

C

Write the date the day after:

1. 2nd February ______
2. 20th June ______
3. 4th August ______
4. 30th November ______
5. 1st April ______
6. 19th January ______
7. 31st December ______
8. 11th May ______
9. 31st March ______
10. 30th July ______

Write in short form.

11. 3rd October 2006 ______
12. 16th May 2010 ______
13. 31st January 2001 ______
14. 10th September 2008 ______
15. 28th June 2020 ______
16. 4th November 2019 ______
17. 13th March 2015 ______
18. 22nd December 2006 ______
19. 30th April 2009 ______
20. 28th July 2003 ______

Score

Unit 8

Strategy

×10 (add 0)
×100 (add 00)
×1000 (add 000)

eg $8 \times 10 = 80$
$8 \times 100 = 800$
$8 \times 1000 = 8000$

1 $9 \times 10 =$ ________
2 $3 \times 100 =$ ________
3 $5 \times 1000 =$ ________
4 $10 \times 16 =$ ________
5 $70 \times 100 =$ ________
6 $294 \times 100 =$ ________
7 $1000 \times 14 =$ ________
8 $56 \times 100 =$ ________
9 $39 \times 10 =$ ________
10 $23 \times 1000 =$ ________
11 $608 \times 10 =$ ________
12 $100 \times 64 =$ ________
13 $109 \times 1000 =$ ________
14 $142 \times 10 =$ ________

Score

Space

Draw and name each special quadrilateral.

1 I have opposite sides equal. All my angles are right angles.
2 I have only one pair of parallel sides.
3 My opposite sides are equal and parallel. My angles are not right angles.
4 I have two pairs of equal adjacent sides.

Problem of the week

Nikko eats baked beans daily. On Monday he ate 40, Tuesday 50, Wednesday 70 and Thursday 100. If he continues in the pattern:

1 how many will he eat on Sunday? ________
2 how many will he eat in the week? ________

Unit 9 Revision

A

1. $7 \times 8 =$ ______
2. $11 + 9 =$ ______
3. $81 \div 9 =$ ______
4. $9 + 7 + 12 =$ ______
5. $9 \times 6 =$ ______
6. $16 - 7 =$ ______
7. $42 \div 6 =$ ______
8. $18 - 7 - 9 =$ ______
9. $6 + 8 + 13 =$ ______
10. $7 \div 7 =$ ______
11. $13 + 8 =$ ______
12. $49 + 4 =$ ______
13. $8 \times 8 =$ ______
14. $100 \div 10 =$ ______
15. $19 + 5 =$ ______
16. $7 \times 0 =$ ______
17. $5 \times 7 =$ ______
18. $64 - 19 =$ ______
19. $30 - 14 - 8 =$ ______
20. $49 \div 7 =$ ______

Score

B

1. value of 5 in 35 601 ______
2. 35c + 80c = ______
3. $\frac{7}{10}$ as a decimal ______
4. True or false. $\frac{1}{4} < \frac{1}{8}$ ______
5. $\frac{1}{8} + \frac{3}{8} =$ ______
6. $\frac{1}{10}$ of a decade ______
7. $2.55 + 90c = ______
8. value of 2 in 19 023 ______
9. 238 cm = ______ m
10. $\frac{1}{4} + \frac{3}{4} =$ ______
11. circle the smallest 5031 3510 5103
12. next multiple of 7 after 56 ______
13. 75 mm = ______ cm
14. $\frac{10}{12} - \frac{7}{12} =$ ______
15. 0·6 as a fraction ______
16. $5 – $2.30 = ______
17. $1.05 + $1.95 = ______
18. factors of 16 ______
19. True or false. $54 \div 9 = 32 - 26$ ______
20. $\frac{1}{3}$ of June ______

Score

C

1. days in 7 weeks ______
2. months in Spring ______
3. hours in 3 days ______
4. years in 2 decades ______
5. 50% of 1 year ______
6. $\frac{1}{5}$ of 35 apples ______
7. 20 minutes after 2:55 ______
8. 12, 19, ______, 33
9. prime number after 33 ______
10. 5000 m = ______ km
11. change from $10 for $3.80 ______
12. perimeter of square with side 11 m ______
13. seconds in $2\frac{1}{2}$ minutes ______
14. centimetres in $5\frac{1}{2}$ metres ______
15. degrees in 3 right angles ______
16. 20 minutes before 11:10 ______
17. $\frac{1}{6}$ of 2 dozen ______
18. date after 30th April ______
19. 20 minutes after 7:15 pm ______
20. 40 mm = ______ cm

Score

Targeting Mental Maths Year 5 • 978-1-922887-27-6

Revision Unit 9

Strategy

Use the strategies you have learnt.

1. 16 + 17 = ______
2. 35 + 36 = ______
3. 46 + 78 = ______
4. 77 + 19 = ______
5. 17 + 49 = ______
6. 39 + 52 = ______
7. 48 + 37 = ______
8. 45 − 16 = ______
9. 54 − 37 = ______
10. 17 × 4 = ______
11. 33 × 4 = ______
12. 14 × 8 = ______
13. 22 × 8 = ______
14. 19 × 100 = ______
15. 27 × 1000 = ______

Score

Space

Draw.

1 obtuse angle	2 triangular prism	3 trapezium	4 square pyramid

Time

Draw the time.

1

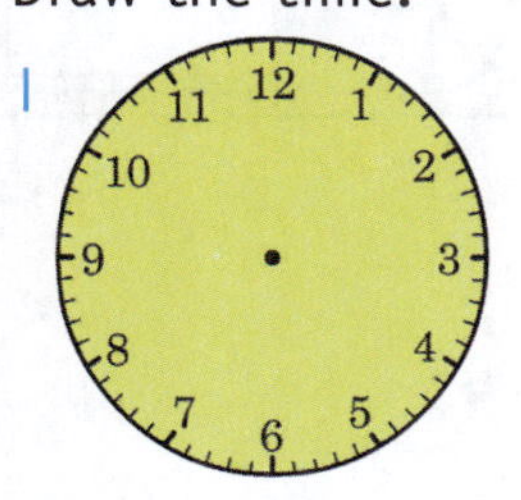

20 to 5

2 :

5 past 11

Fractions

Colour.

1

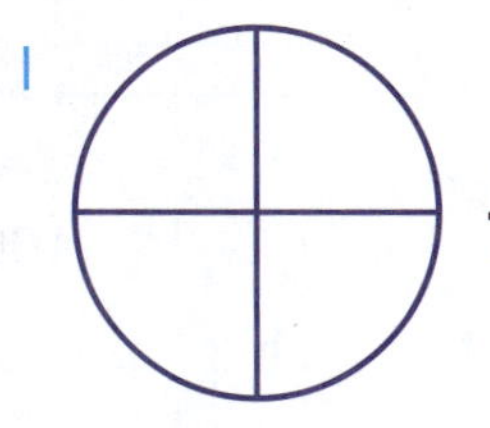

$\frac{3}{4}$

2

$\frac{5}{8}$

Number

Complete.

1

Factor	7		8	9
Factor		5	6	
Multiple	63	45		54

2

Factor	2		7	
Factor	16	8		3
Multiple		88	56	60

Length

Measure in mm.

1

2

3 Draw a line $5\frac{1}{2}$ cm long.

Unit 10

A

1 7 + 9 + 5 = ______
2 6 + 8 + 5 = ______
3 9 + 3 + 8 = ______
4 5 + 9 + 6 = ______
5 6 + 7 + 8 = ______
6 8 + 9 + 5 = ______
7 7 + 7 + 8 = ______
8 7 + 8 + 7 = ______
9 8 + 6 + 9 = ______
10 9 + 2 + 9 = ______
11 8 + 7 + 8 = ______
12 6 + 5 + 6 = ______
13 9 + 7 + 8 = ______
14 7 + 4 + 8 = ______
15 8 + 5 + 9 = ______
16 9 + 4 + 6 = ______
17 6 + 9 + 6 = ______
18 8 + 8 + 4 = ______
19 7 + 3 + 7 = ______
20 9 + 6 + 5 = ______

Score

B Arrange in ascending order.

1 $\frac{1}{2}$, 0·3, 0·01, $\frac{1}{10}$, 0·6 ______
2 0·99, $\frac{1}{4}$, $\frac{1}{5}$, 0·77, 0·19 ______
3 0·70, $\frac{1}{100}$, $\frac{11}{100}$, 0·10, $\frac{100}{100}$ ______
4 2·3, 3·02, 1·97, 2·1 ______

Arrange in descending order.

5 $\frac{1}{2}$, $\frac{1}{8}$, $\frac{1}{3}$, $\frac{1}{5}$, $\frac{1}{4}$ ______
6 0·25, 5·22, 0·52, 2·5, 5·2 ______
7 0·8, 1·8, 1·5, $\frac{1}{4}$, 0·45 ______
8 1·4, 1·14, 4·1, $\frac{1}{4}$ ______

Score

C

1 50 614 = 50 000 + 600 + ______ + 4
2 72 009 = ______ + 2000 + 9
3 ______ = 20 000 + 4000 + 6
4 41 300 = ______ + ______ + ______
5 value of 7 in 37 203 ______
6 value of 1 in 29 014 ______
7 297c = $ ______
8 $11.04 = ______ c
9 $8.40 = ______ c
10 3015c = $ ______

Use >, <, or =.

11 1 g ______ 1000 kg
12 2 cm ______ 20 mm
13 72 min ______ $1\frac{1}{2}$ hr
14 100 mL ______ $\frac{1}{10}$ L
15 $\frac{1}{2}$ km ______ 250 m
16 300 cm ______ 3 m
17 150 sec ______ $2\frac{1}{2}$ min
18 30 yr ______ 2 decades
19 250 yr ______ $\frac{1}{4}$ millenium
20 9·7 cm ______ 90 mm

Score

Unit 10

Strategy

× by 10, × by 5

× by ten – add a zero
eg 17 × 10 = 170

× by 5 – add a zero and halve
eg 17 × 5 = 170 ÷ 2
= 85

1 15 × 10 = ______

2 83 × 10 = ______

3 715 × 10 = ______

4 14 × 5 = ______

5 26 × 5 = ______

6 31 × 5 = ______

7 52 × 5 = ______

8 74 × 5 = ______

9 110 × 5 = ______

10 282 × 5 = ______

Score

Data

Day	Tally	Total
1	𝍸 III	
2	𝍸 𝍸 𝍸 IIII	
3	IIII	
4	𝍸 𝍸 𝍸 𝍸 II	
5	𝍸 𝍸 I	

1 What could this data represent?

______

2 Draw a graph to show the data.

3 Give the graph a title. 4 Label the axes.

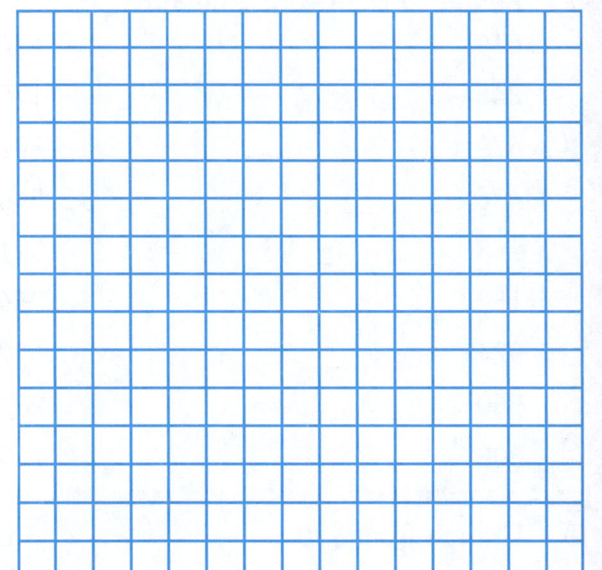

Problem of the week

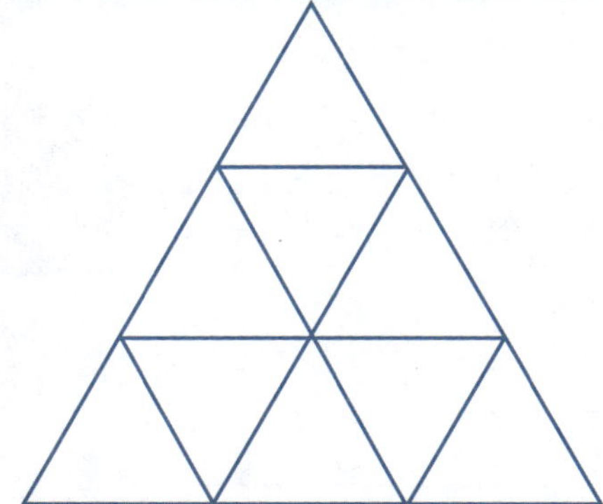

How many parallograms can you find?

1 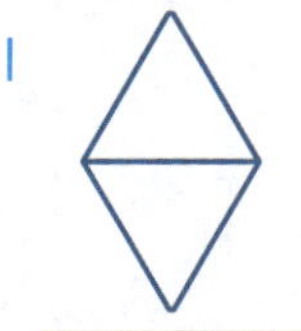

2 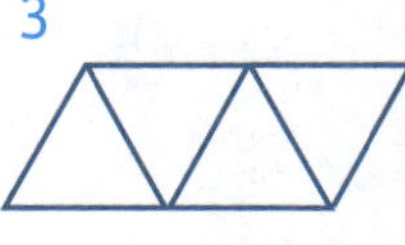

3 

4 Total

Unit 11

A

1 $3 \times 7 + 8 =$ ______

2 $6 \times 8 + 11 =$ ______

3 $5 \times 7 + 5 =$ ______

4 $4 \times 9 + 7 =$ ______

5 $8 \times 6 + 9 =$ ______

6 $7 \times 7 + 11 =$ ______

7 $4 \times 5 + 13 =$ ______

8 $8 \times 9 + 6 =$ ______

9 $5 \times 5 + 8 =$ ______

10 $7 \times 9 + 12 =$ ______

11 $3 \times 8 + 9 =$ ______

12 $6 \times 9 + 4 =$ ______

13 $9 \times 7 + 6 =$ ______

14 $4 \times 6 + 7 =$ ______

15 $4 \times 7 + 9 =$ ______

16 $7 \times 8 + 12 =$ ______

17 $3 \times 9 + 11 =$ ______

18 $8 \times 8 + 7 =$ ______

19 $6 \times 7 + 5 =$ ______

20 $4 \times 8 + 9 =$ ______

Score

B

Round to the nearest hundred.

1 619 ______

2 245 ______

3 879 ______

4 436 ______

5 783 ______

6 157 ______

7 924 ______

8 560 ______

9 308 ______

Round to the nearest thousand.

10 4301 ______

11 1753 ______

12 8029 ______

13 5462 ______

14 511 ______

15 954 ______

16 6274 ______

17 2827 ______

18 7136 ______

Score

C

Change from $20.

1 $7.55 ______

2 $13.05 ______

3 $9.25 ______

4 $17.70 ______

5 $15.45 ______

6 $8.15 ______

7 $16.65 ______

8 $19.20 ______

9 $12.55 ______

10 $11.30 ______

3 coins to make:

11 65c ______

12 35c ______

13 $2.05 ______

14 $2 ______

15 $4.50 ______

16 $1.15 ______

17 20c ______

18 $3 ______

19 $4.05 ______

20 $0.75 ______

Score

Targeting Mental Maths Year 5 • 978-1-922887-27-6

Strategy

Jump strategy

eg 96 + 57 = 96 + 50 + 7
= 153
81 − 36 = 81 − 30 − 6
= 45

1 58 + 67 = ____________
= ________

2 94 + 89 = ____________
= ________

3 81 + 47 = ____________
= ________

4 137 + 66 = ____________
= ________

5 91 − 54 = ____________
= ________

6 73 − 28 = ____________
= ________

7 82 − 69 = ____________
= ________

8 120 − 75 = ____________
= ________

9 165 − 87 = ____________
= ________

Score

Space

Name these shapes.

1 

2 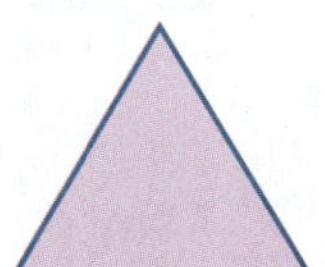

3 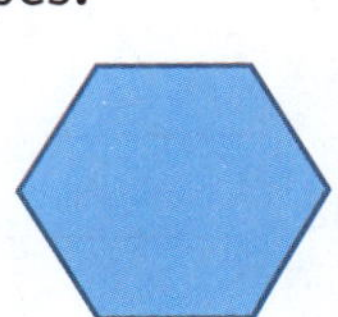

4 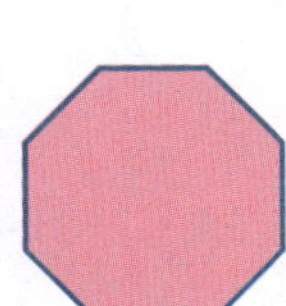

5 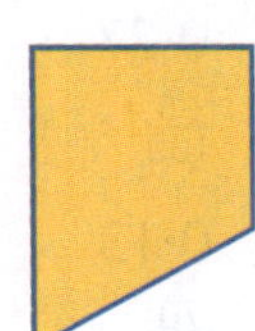

________ ________ ________ ________ ________

Number

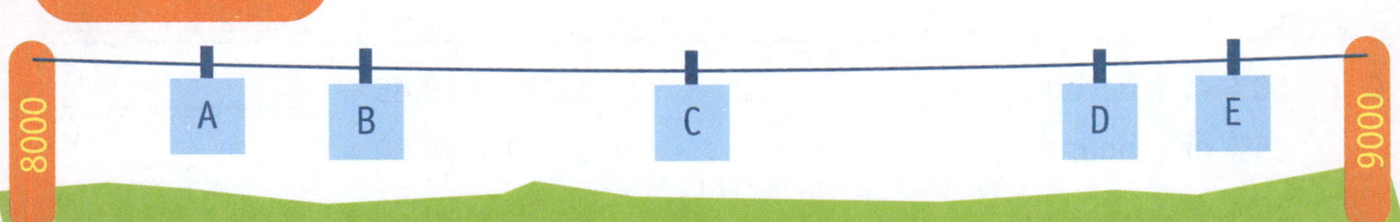

What is the approximate value of each card?

A ________ B ________ C ________ D ________ E ________

Problem of the week

Gran's clock gains 15 minutes every hour. She sets it exactly at midday. After dinner her watch reads 7:20.

What is the time on her clock? ____________

Unit 12

A

1 10 × 9 = ______
2 7 × 6 = ______
3 4 × 5 = ______
4 7 × 9 = ______
5 0 × 8 = ______
6 6 × 9 = ______
7 4 × 8 = ______
8 6 × 6 = ______
9 3 × 7 = ______
10 7 × 5 = ______
11 8 × 8 = ______
12 10 × 6 = ______
13 7 × 8 = ______
14 4 × 7 = ______
15 4 × 4 = ______
16 9 × 9 = ______
17 5 × 9 = ______
18 10 × 0 = ______
19 6 × 7 = ______
20 3 × 9 = ______

Score

B Write decimals as fractions and fractions as decimals.

1 $\frac{16}{100}$ ______
2 0·32 ______
3 $\frac{84}{100}$ ______
4 0·27 ______
5 $\frac{56}{100}$ ______
6 0·13 ______
7 $\frac{70}{100}$ ______
8 0·08 ______
9 0·99 ______
10 $\frac{4}{100}$ ______
11 $\frac{47}{100}$ ______
12 $\frac{28}{100}$ ______
13 0·68 ______
14 0·11 ______
15 $\frac{1}{100}$ ______
16 0·82 ______
17 $\frac{59}{100}$ ______
18 $\frac{91}{100}$ ______
19 0·37 ______
20 $\frac{29}{100}$ ______

Score

C

1 50% of 1 decade ______
2 25% of 1 hour ______
3 20% of $10 ______
4 75% of 1 century ______
5 10% of $100 ______
6 25% of 1 dozen ______
7 20% of 10 minutes ______
8 50% of 1 km ______
9 75% of 1 year ______
10 100% of $2 ______
11 $2.15 + $3.85 = ______
12 $1.60 + $4.80 = ______
13 $9.50 + $6.20 = ______
14 $3.45 + $8.65 = ______
15 $1.55 + $2.30 = ______
16 $8.05 + $7.90 = ______
17 $6.30 + $9.80 = ______
18 $2.10 + $12.45 = ______
19 $3.85 + $7.20 = ______
20 $7.75 + $11.25 = ______

Score

Strategy

÷ by 10

÷ by 5

÷ by 10 – last number is the remainder

eg $149 \div 10 = 14\frac{9}{10}$

÷ by 5 – ÷ by 10 and double

eg $83 \div 5 = 8\frac{3}{10} \times 2$

$= 16\frac{6}{10}$

1 57 ÷ 10 = ________

2 93 ÷ 10 = ________

3 128 ÷ 10 = ________

4 206 ÷ 10 = ________

5 72 ÷ 5 = ________

= ________

6 101 ÷ 5 = ________

= ________

7 94 ÷ 5 = ________

= ________

8 233 ÷ 5 = ________

= ________

Score

Number

Divide the jelly beans.

5 ÷ 4 ________

1 ÷ 8 ________

2 ÷ 5 ________

6 ÷ 7 ________

7 ÷ 3 ________

3 ÷ 6 ________

4 ÷ 1 ________

8 ÷ 9 ________

Hundredths

Write each as a fraction and as a decimal.

1

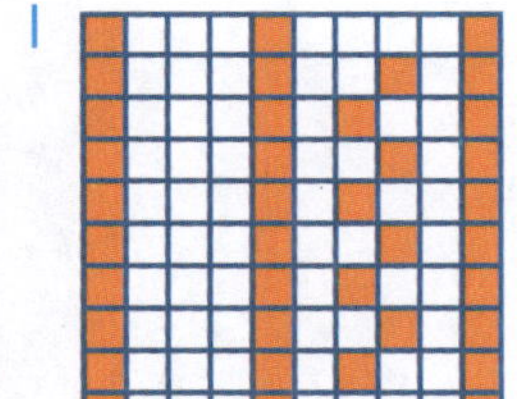

2

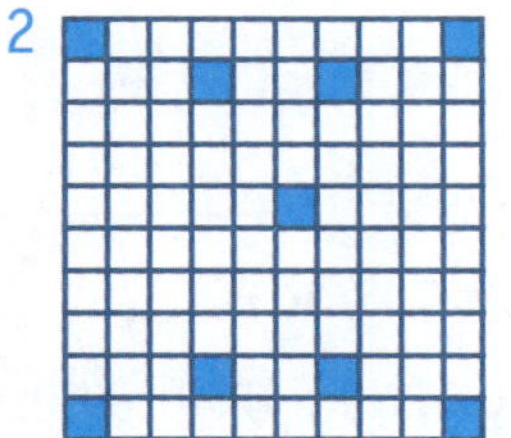

________ ________ ________ ________

Problem of the week

Didi and Davo work at the dog parlour. Didi takes 20 minutes to wash a dog and Davo takes 30 minutes. They make the same amount of money in an 8 hour day. Didi charges $20 per wash.

How much does Davo charge?

____

Unit 13

A

1 81 ÷ 9 = ______
2 42 ÷ 7 = ______
3 64 ÷ 8 = ______
4 35 ÷ 7 = ______
5 54 ÷ 9 = ______
6 48 ÷ 8 = ______
7 9 ÷ 9 = ______
8 45 ÷ 9 = ______
9 56 ÷ 8 = ______
10 27 ÷ 9 = ______
11 49 ÷ 7 = ______
12 72 ÷ 9 = ______
13 54 ÷ 6 = ______
14 63 ÷ 7 = ______
15 72 ÷ 8 = ______
16 15 ÷ 5 = ______
17 63 ÷ 9 = ______
18 36 ÷ 6 = ______
19 32 ÷ 8 = ______
20 36 ÷ 9 = ______

Score

B

Write the factors of:

1 12 ______
2 20 ______
3 15 ______
4 8 ______
5 27 ______
6 36 ______
7 29 ______
8 50 ______

Write the next multiple of:

9 7 after 10 ______
10 12 after 0 ______
11 4 after 30 ______
12 9 after 50 ______
13 3 after 30 ______
14 6 after 40 ______
15 8 after 60 ______
16 20 after 40 ______

Score

C

1 days in a leap year ______
2 weeks in 2 fortnights ______
3 days in September ______
4 month before January ______
5 months in Winter ______
6 minutes in $1\frac{1}{4}$ hours ______
7 seconds in $4\frac{1}{2}$ minutes ______
8 days in 12 weeks ______
9 years in 8·5 decades ______
10 seconds in $\frac{1}{4}$ minute ______
11 total of 17, 14, 8 ______
12 product of 10 and 16 ______
13 difference between 3 and 21 ______
14 8 times 20 ______
15 sum of 43 and 28 ______
16 take 29 from 51 ______
17 11 lots of 7 ______
18 55 minus 38 ______
19 subtract 17 from 41 ______
20 How many 4s in 48? ______

Score

Strategy
÷ by 4

halve and halve again
68 ÷ 4 (= 68 ÷ 2 ÷ 2)
= 34 ÷ 2
= 17

1 52 ÷ 4 = ______
= ______

2 88 ÷ 4 = ______
= ______

3 64 ÷ 4 = ______
= ______

4 96 ÷ 4 = ______
= ______

5 128 ÷ 4 = ______
= ______

6 260 ÷ 4 = ______
= ______

7 74 ÷ 4 = ______
= ______

8 152 ÷ 4 = ______
= ______

Score

Patterns

Jemima loved reading. The first day of her holiday she read 23 pages. Each day after that she read 3 more pages than the day before.

How long did it take to read her 224 page book? ______

Day	1	2	3						
Pages	23	26	29						
Total		49							

Chance

What is the chance?

1 You are a cat. ______

2 You will be older next year. ______

3 The sun will shine for 24 hours. ______

4 The family next door will move. ______

5 You will win a raffle. ______

Problem of the week

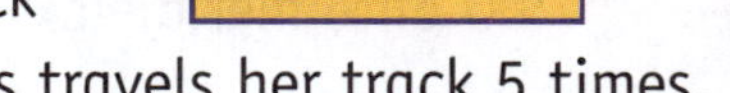

Mac's car set has a track 3·24 m long. Marnie's set has a track 4·08 m long. Mac's car travels the track 6 times and Marnie's travels her track 5 times.

Whose car travels furthest? ______ By how much? ______

Unit 14

A

1. 0·9 + 0·3 = ________
2. 0·8 + 0·4 = ________
3. 0·5 + 0·5 = ________
4. 0·3 + 0·6 = ________
5. 0·2 + 0·9 = ________
6. 0·7 + 0·6 = ________
7. 0·5 + 0·8 = ________
8. 0·8 − 0·2 = ________
9. 0·5 − 0·4 = ________
10. 0·9 − 0·6 = ________
11. 1·1 − 0·5 = ________
12. 1·3 − 0·9 = ________
13. 1·8 − 0·8 = ________
14. 1·6 − 0·9 = ________
15. 3·6 + 2·7 = ________
16. 5·8 + 6·9 = ________
17. 9·9 + 4·4 = ________
18. 6·8 − 2·5 = ________
19. 8·7 − 3·1 = ________
20. 9·2 − 5·6 = ________

Score

B

1. 5, 9, 13, 17, ________
2. 82, 73, 64, 55, ________
3. 1, 2, 4, 8, ________
4. 32, 37, ________, 47, 52
5. 60, 53, 46, ________, 32
6. 81, 27, 9, 3, ________
7. ________, 4, 9, 16, 25
8. 1, 4, 16, ________, 256
9. 6, ________, 28, 39, 50
10. 1, 2, 4, 7, 11, ________
11. 50% of 10 = ________
12. 10% of 80 = ________
13. 25% of 60 = ________
14. 75% of 12 = ________
15. 20% of 45 = ________
16. 50% of $5 = ________
17. 75% of $20 = ________
18. 20% of $15 = ________
19. 25% of $48 = ________
20. 10% of $3 = ________

Score

C

Write using abbreviations.

1. 3 millilitres ________
2. 8 centimetres ________
3. 29 grams ________
4. 512 kilograms ________
5. 76 kilograms ________
6. 115 litres ________
7. 35 millimetres ________
8. 7:00 ante meridiem ________
9. 612 metres ________
10. 9:30 post meridiem ________

Complete:

11. 5·45 m = ________ cm
12. 3·25 cm = ________ mm
13. 11·6 kg = ________ g
14. 8·04 km = ________ m
15. 17·25 hr = ________ min
16. 3090 m = ________ km
17. 854 mm = ________ cm
18. 6430 g = ________ kg
19. 400 min = ________ hr
20. 870 cm = ________ m

Score

Unit 14

Strategy
÷ by 8

halve, halve and halve again
eg 96 ÷ 8 (= 96 ÷ 2 ÷ 2 ÷ 2)
= 48 ÷ 2 ÷ 2
= 24 ÷ 2
= 12

1 160 ÷ 8 = 80 ÷ 2 ÷ 2
= 40 ÷ 2
= ______

2 144 ÷ 8 = ______
= ______
= ______

3 112 ÷ 8 = ______
= ______
= ______

4 192 ÷ 8 = ______
= ______
= ______

5 200 ÷ 8 = ______
= ______
= ______

6 176 ÷ 8 = ______
= ______
= ______

Score

Volume

The models are built with blocks. How many blocks are needed for each model?

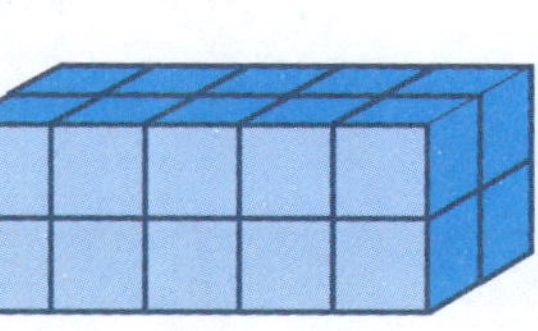

1 ______

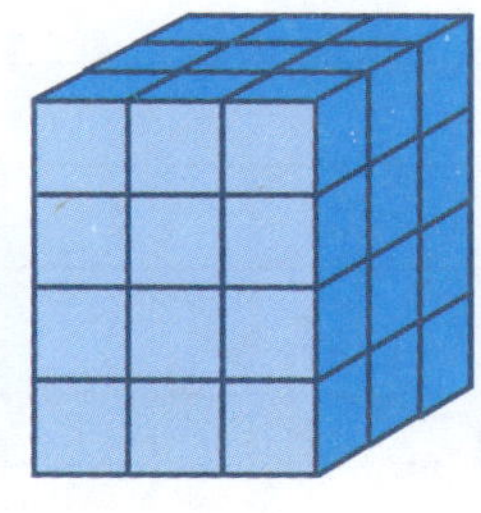

2 ______

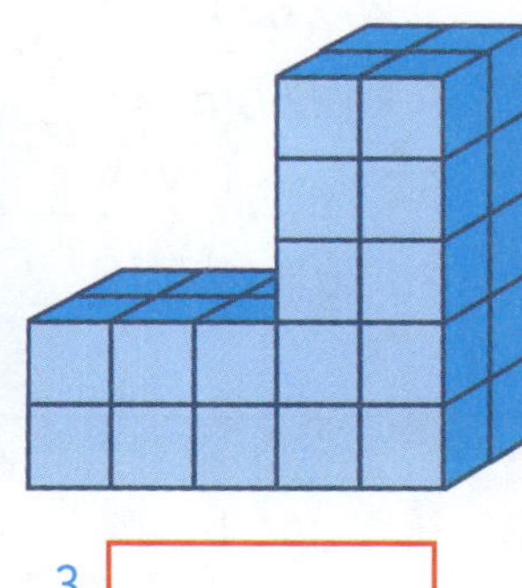

3 ______

Space

In the world around you where do you see:

1 cones? ______

2 cylinders? ______

3 spheres? ______

Problem of the week

Merlin has lost his magic number. Help him find it. Firstly it was tripled. Then it was halved. Then 15 was subtracted from it to get an answer of 21.

What is it? ______

Unit 15

A

1 100 ÷ 10 = ________
2 9 ÷ 3 = ________
3 6 × 7 = ________
4 24 ÷ 4 = ________
5 8 ÷ 8 = ________
6 56 ÷ 7 = ________
7 9 × 6 = ________
8 8 × 8 = ________
9 24 ÷ 8 = ________
10 4 × 7 = ________
11 25 ÷ 5 = ________
12 7 × 1 = ________
13 18 ÷ 9 = ________
14 7 × 8 = ________
15 49 ÷ 7 = ________
16 6 × 0 = ________
17 45 ÷ 5 = ________
18 9 × 3 = ________
19 48 ÷ 6 = ________
20 7 × 9 = ________

Score

B

Change to grams.

1 3 kg ________
2 $4\frac{1}{2}$ kg ________
3 13 kg ________
4 $7\frac{1}{4}$ kg ________
5 1 kg 700 g ________
6 9 kg 550 g ________
7 $5\frac{3}{4}$ kg ________
8 26 kg 118 g ________

Change to kilograms and grams.

9 7000 g ________
10 12 000 g ________
11 16 000 g ________
12 8500 g ________
13 10 750 g ________
14 43 100 g ________
15 82 745 g ________
16 2008 g ________

Score

C

Change to 24-hour time.

1 6:00 am ________
2 9:00 pm ________
3 6:00 pm ________
4 11:00 am ________
5 10:00 pm ________
6 8:00 pm ________
7 7:15 am ________
8 1:30 pm ________
9 5:40 pm ________
10 9:50 am ________

Change to 12-hour time (am and pm).

11 05:00 ________
12 14:00 ________
13 23:00 ________
14 01:00 ________
15 10:00 ________
16 16:00 ________
17 20:15 ________
18 19:30 ________
19 18:05 ________
20 03:10 ________

Score

Targeting Mental Maths Year 5 • 978-1-922887-27-6

Unit 15

Strategy
Add to tens

eg 16 + 27 + 24 = 40 + 27
= 67

1 15 + 23 + 15 = ______
= ______

2 35 + 17 + 13 = ______
= ______

3 19 + 11 + 18 = ______
= ______

4 8 + 59 + 12 = ______
= ______

5 17 + 7 + 13 = ______
= ______

6 21 + 46 + 19 = ______
= ______

7 25 + 15 + 36 = ______
= ______

8 83 + 14 + 6 = ______
= ______

9 53 + 69 + 47 = ______
= ______

Score

Space

Which 3D shape has a net with:

1 One square face and four triangular faces?

______

2 Two triangular faces and three rectangular faces?

______

3 Three square faces and two triangular faces?

______

Position

Write in the eight main compass points.

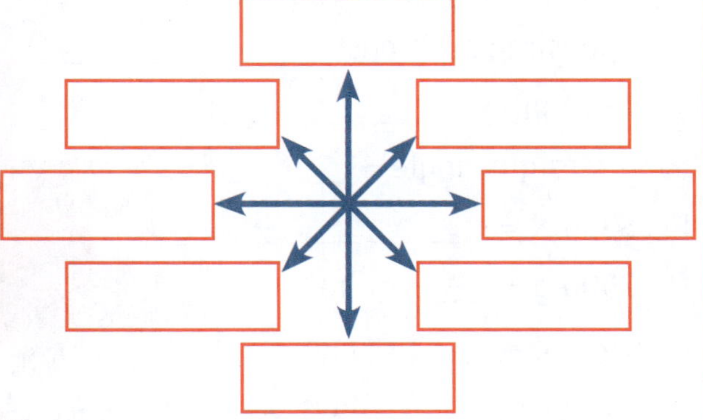

Problem of the week

27 small cubes make this block. The block is painted yellow.

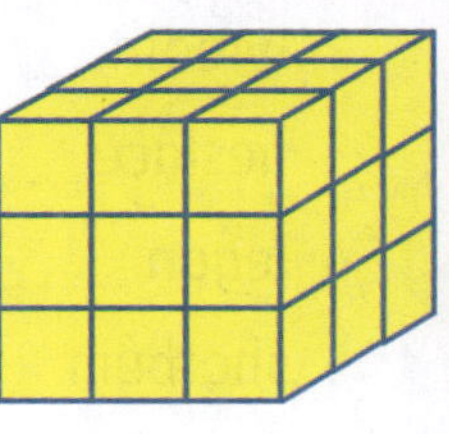

How many cubes are painted:

on one side? ______

on two sides? ______

on three sides? ______

on no sides? ______

Unit 16

A

1 6 + 9 = ______
2 16 + 19 = ______
3 36 + 49 = ______
4 8 + 5 = ______
5 18 + 15 = ______
6 58 + 25 = ______
7 7 + 9 = ______
8 17 + 19 = ______
9 27 + 59 = ______
10 4 + 8 = ______
11 14 + 18 = ______
12 34 + 38 = ______
13 2 + 9 = ______
14 12 + 19 = ______
15 42 + 39 = ______
16 6 + 7 = ______
17 16 + 17 = ______
18 76 + 17 = ______
19 3 + 8 = ______
20 43 + 38 = ______

Score

B Write as a fraction.

1 0·3 ______
2 0·6 ______
3 0·8 ______
4 0·13 ______
5 0·72 ______
6 0·95 ______
7 0·29 ______
8 1·2 ______
9 1·5 ______
10 2·28 ______
11 0·64 ______
12 3·73 ______
13 0·11 ______
14 0·36 ______
15 5·55 ______
16 0·99 ______
17 1·01 ______
18 0·4 ______

Score

C Unscramble these 'space' words.

1 aceongd ______
2 lyredinc ______
3 neslace ______
4 legan ______
5 shorbum ______
6 sitonoip ______
7 uboset ______
8 mipardy ______
9 seeloscis ______
10 quietrealal ______

Complete:

11 opposite north ______
12 opposite east ______
13 opposite south-west ______
14 opposite south-east ______
15 1250 mL = ______ L
16 a straight angle = ______ °
17 $375 K = ______
18 2500 g = ______ kg
19 1·8 cm = ______ mm
20 spring = ______ days

Score

Strategy
Associative law

Two or more numbers can be added in any order.
eg 8 + 7 + 5 = 8 + 5 + 7 or 5 + 8 + 7 or 7 + 5 + 8

Write two different ways to add each one.

1 9 + 6 + 11 = ______________ or ______________

2 15 + 8 + 6 = ______________ or ______________

3 19 + 7 + 12 = ______________ or ______________

4 21 + 16 + 19 = ______________ or ______________

5 36 + 7 + 15 = ______________ or ______________

6 62 + 14 + 28 = ______________ or ______________

Score

Space

Name and draw the 3D objects you can make from these nets.

1

Draw

____

2

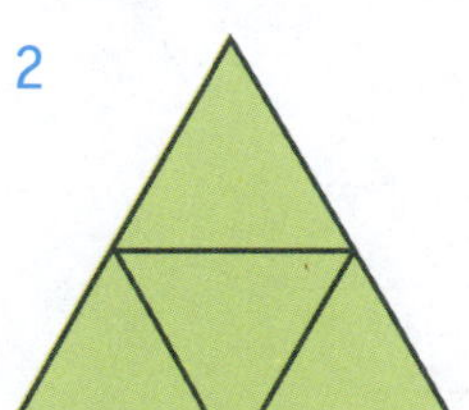

Draw

____

Problem of the week

These three counters are to be placed in the boxes.

The blue one has to be to the left of the others. The green one is in the middle of the three and the red one is to the right. They cannot be next to each other.

In how many different ways can you arrange them? ______________

A	B	C	D	E	F

Unit 17

A

1 6 × 10 = ________
2 19 × 10 = ________
3 204 × 10 = ________
4 376 × 10 = ________
5 8 × 20 = ________
6 14 × 20 = ________
7 28 × 20 = ________
8 103 × 20 = ________
9 69 × 20 = ________
10 215 × 20 = ________
11 4 × 30 = ________
12 7 × 30 = ________
13 9 × 40 = ________
14 11 × 40 = ________
15 8 × 50 = ________
16 2 × 60 = ________
17 6 × 70 = ________
18 9 × 80 = ________
19 7 × 50 = ________
20 10 × 90 = ________

Score

B

Position:

1 after 3rd ________
2 after 11th ________
3 before 9th ________
4 before 3rd ________
5 after tied 7th ________
6 after tied 10th ________
7 before tied 6th ________
8 after 20th ________

Write three multiples for each factor.

9 3 ________
10 7 ________
11 9 ________

Write three factors for each multiple.

12 45 ________
13 60 ________
14 42 ________

Score

C

Area and perimeter.

1

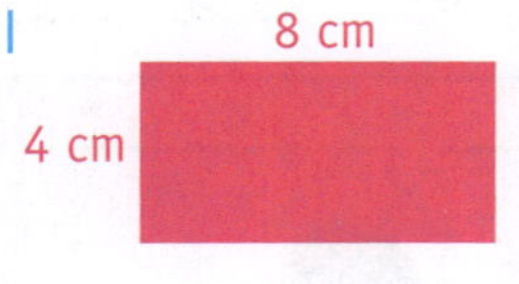

P = ________
A = ________

2 12 cm, 12 cm

P = ________
A = ________

3 7 cm, 2 cm

P = ________
A = ________

4 9 cm, 3 cm

P = ________
A = ________

What am I?

5 I have one curved surface and one flat surface. ________
6 I have two triangular faces and three rectangular faces. ________
7 I only have one curved surface. ________
8 I have four triangular faces. ________
9 I have two circular ends and one curved surface. ________

Score

Targeting Mental Maths Year 5 • 978-1-922887-27-6

Unit 1

A
1 0 2 21 3 63 4 25
5 81 6 64 7 24 8 49
9 48 10 40 11 72 12 56
13 54 14 36 15 42 16 72
17 63 18 45 19 32 20 48

B
1 0.6 2 0.006 3 2
4 70 5 0.07 6 0.06
7 20 8 0.6 9 0.002
10 18 058 11 77 370 12 73 370
13 11 028 14 7500 15 92 604
16 51 009 17 2003

C
1 acute 2 right angle 3 obtuse
4 obtuse 5 acute 6 obtuse 7 obtuse
8 acute 9 obtuse 10 acute 11 35
12 60 13 3 14 30 15 366
16 210 17 31 18 96 19 10
20 1000

Strategy
1 15 + 15 + 1 = 31 2 20 + 20 − 1 = 39
3 14 + 14 − 1 = 27 4 18 + 18 + 1 = 37
5 25 + 25 + 1 = 51 6 32 + 32 + 1 = 65
7 36 + 36 − 1 = 71 8 27 + 27 − 1 = 53
9 34 + 34 − 1 = 67
10 45 + 45 + 1 = 91

Number
1 3165 2 5583
3 4

Space
1 & 6 red 2 & 4 blue 3 & 5 green

Problem of the week 64 and 98

Unit 2

A
1 8 2 6 3 5 4 8
5 10 6 6 7 7 8 5
9 9 10 1 11 6 12 6
13 9 14 7 15 4 16 6
17 6 18 4 19 10 20 1

B
1 53 2 156 3 504 4 207
5 123 6 215 7 139 8 197

C
1 16 cm 2 36 m 3 24 km
4 48 mm 5 144 cm 6 22 m
7 44 mm 8 46 km 9 52 cm
10 88 m 11 $8.60 12 $3.20
13 $0.50 14 $5.80 15 $6.85
16 $8.25 17 $4.65 18 $1.75
19 5c 20 $2.35

Strategy
1 40 + 80 + 6 + 5 = 131
2 90 + 20 + 3 + 9 = 122
3 60 + 50 + 8 + 6 = 124
4 50 + 80 + 7 + 6 = 143
5 70 + 20 + 7 + 9 = 106
6 60 + 20 + 9 + 4 = 93
7 80 + 70 + 5 + 4 = 159
8 90 + 40 + 5 + 8 = 143

Time
1 3:40 2 3 8:05 4
5 1:20

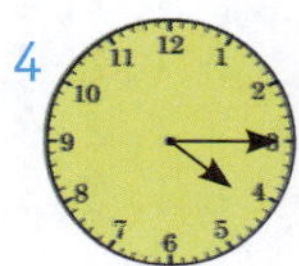

Chance Teacher check

Problem of the week 12

Unit 3

A
1 12 2 17 3 13 4 20
5 11 6 20 7 14 8 11
9 21 10 13 11 9 12 15
13 15 14 13 15 18 16 12
17 25 18 15 19 14 20 24

B
1 $1.15 2 $1.15 3 $1.45 4 $1.55
5 $1.05 6 $1.80 7 $2.25 8 $2.65
9 $3.20 Estimates may vary. 10 110
11 120 12 150 13 170 14 250
15 290 16 230 17 240 18 330

C
1 6 months 2 6 3 $1.25
4 $1.20 5 15 minutes 6 10 yr
7 1 week 8 12 9 15 seconds
10 759 cm 11 63 12 72
13 240 14 150 15 250
16 750 17 45 000 18 8 500
19 450° 20 925c

Strategy
1 27 + 40 − 1 = 66 2 54 + 60 + 1 = 115
3 38 + 60 − 2 = 96 4 67 + 20 + 2 = 89
5 98 + 80 − 2 = 176 6 29 + 50 + 2 = 81
7 55 + 70 + 1 = 126 8 88 + 70 − 1 = 157
9 108 + 70 + 3 = 181

Answers

Number 55, 22, 121, 47, 135, 39
Money 1 $586 2 $14
Problem of the week 78

Unit 4

A
1 12 2 5 3 10 4 9
5 11 6 5 7 9 8 11
9 8 10 0 11 9 12 8
13 5 14 9 15 12 16 7
17 6 18 7 19 6 20 14

B
1 1, 2, 4, 8, 16
2 1, 2, 3, 4, 6, 8, 12, 24
3 1, 3, 7, 9, 21, 63
4 1, 2, 4, 7, 8, 14, 28, 56
5 1, 2, 4, 7, 14, 28
6 1, 2, 3, 6, 7, 14, 21, 42
7 1, 2, 3, 4, 6, 9, 12, 18, 36
8 1, 2, 3, 4, 6, 8, 9, 12, 18, 24, 36, 72
9 24 10 56 11 21 12 63
13 54 14 75 15 42 16 32

C
1 $20 + $10 + $5 2 $10 + $10 + $5
3 $50 + $20 + $5 4 $10 + $5 + $5
5 $50 + $20 + $20 6 $100 + $5 + $5
7 $100 + $50 + $5 8 $100 + $10 + $5
9 $100 + $100 + $20
10 $50 + $10 + $5 11 4:50 12 2:35
13 3:40 14 7:10 15 2:05 16 5:55
17 10:30 18 9:05 19 12:25 20 2:45

Strategy
1 48 + 30 + 7 = 85 2 95 + 60 + 6 = 161
3 88 + 50 + 4 = 142 4 53 + 70 + 7 = 130
5 66 + 40 + 7 = 113 6 81 − 20 − 4 = 57
7 92 − 60 − 5 = 27 8 64 − 30 − 7 = 27
9 73 − 20 − 8 = 45

Fractions Teacher check
Space Teacher check
Number

Factor	6	7	7	8	6	9
Factor	9	8	3	8	7	4
Product	54	56	21	64	42	36

Problem of the week 60

Unit 5

A
1 42 2 7 3 49 4 8
5 48 6 56 7 7 8 56
9 10 10 7 11 63 12 54
13 8 14 36 15 45 16 6
17 0 18 64 19 6 20 27

B
1 0·3 2 0·5 3 0·2 4 0·7
5 0·9 6 $\frac{4}{10}$ 7 $\frac{8}{10}$ 8 $\frac{3}{10}$
9 $\frac{6}{10}$ 10 = 11 > 12 >
13 > 14 > 15 = 16 >
17 = 18 <

C
1 41 2 81 3 18 4 80
5 $\frac{1}{6}$ 6 0·65 7 20 8 5
9 10 10 9 11-16 Teacher check
17 3 18 8 19 8 20 6

Strategy
1 16 × 2 × 2 = 64 2 13 × 2 × 2 = 52
3 15 × 2 × 2 = 60 4 19 × 2 × 2 = 76
5 22 × 2 × 2 = 88 6 50 × 2 × 2 = 200
7 110 × 2 × 2 = 440 8 61 × 2 × 2 = 244
9 32 × 2 × 2 = 128 10 75 × 2 × 2 = 300

Patterns
1

1	2	3	4	5	6	7	8
8	15	22	29	36	43	50	57

2 shapes × 7 + 1

Angles
1 acute 2 obtuse 3 right angle
4 acute

Problem of the week
1 11:42 pm 2 31st December

Unit 6

A
1 22 2 15 3 31 4 7
5 23 6 44 7 62 8 32
9 25 10 13 11 43 12 21
13 75 14 5 15 32 16 43
17 27 18 46 19 24 20 53

B
1 = 2 > 3 > 4 <
5 = 6 < 7 < 8 <
9 < 10 > 11 = 12 =
13 > 14 < 15 > 16 <
17 > 18 =

C
1 11 | 2 13 | 3 29 | 4 37
5 3 | 6 1500 | 7 9 | 8 450
9 4 | 10 35 | 11 L | 12 km
13 m/cm | 14 kg | 15 hours | 16 cm
17 m | 18 grams | 19 L | 20 t/kg

Strategy
1 128 | 2 104 | 3 256 | 4 144
5 200 | 6 248 | 7 152 | 8 336
9 120 | 10 408

Length
1 45 mm | 2 73 mm | 3 80 mm | 4 44 mm
5-8 Teacher check

Area Teacher check

Problem of the week 5 days

Unit 7

A
1 20 | 2 20 | 3 20 | 4 20
5 20 | 6 18 | 7 25 | 8 28
9 23 | 10 22 | 11 22 | 12 20
13 21 | 14 21 | 15 24 | 16 23
17 20 | 18 25 | 19 25 | 20 24

B
1 1·5 | 2 1·7 | 3 6·3 | 4 2·21
5 8·1 | 6 8·04 | 7 11·3 | 8 0·35
9 36 | 10 89 | 11 274 | 12 327
13 584 | 14 906 | 15 53 | 16 2

C
1 T | 2 F | 3 T | 4 F
5 T | 6 T | 7 T | 8 F
9 F | 10 T | 11 right angle
12 acute | 13 acute | 14 obtuse
15 obtuse | 16 acute | 17 acute
18 acute | 19 obtuse | 20 obtuse

Strategy
1 13, 23, 53 | 2 13, 23, 93
3 12, 22, 72 | 4 11, 21, 41
5 17, 27, 97 | 6 14, 24, 84
7 13, 23, 63 | 8 10, 20, 60
9 15, 25, 95

Space Teacher check

Length
1 $1\frac{1}{2}$ cm, 6 cm, $\frac{1}{2}$ cm, 5 cm, 1 cm, 11 cm
2 25 cm

Problem of the week 4 pizzas

Unit 8

A
1 13 | 2 4 | 3 9 | 4 13
5 5 | 6 10 | 7 10 | 8 3
9 8 | 10 6 | 11 9 | 12 0
13 7 | 14 3 | 15 9 | 16 2
17 3 | 18 9 | 19 9 | 20 9

B
1 $\frac{2}{4}$ ($\frac{1}{2}$) | 2 1 | 3 $\frac{4}{5}$ | 4 $\frac{7}{8}$
5 $\frac{7}{12}$ | 6 $\frac{2}{4}$ ($\frac{1}{2}$) | 7 $\frac{7}{12}$ | 8 $\frac{1}{3}$
9 $\frac{2}{5}$ | 10 3 apples | 11 10c | 12 20c
13 4 books | 14 12 sweets | 15 1 egg | 16 $25
17 7 bananas | 18 20 yrs

C
1 3rd February | 2 21st June
3 5th August | 4 1st December
5 2nd April | 6 20th January
7 1st January | 8 12th May
9 1st April | 10 31st July
11 3/10/06 | 12 16/5/10
13 31/1/01 | 14 10/9/08
15 28/6/20 | 16 4/11/19
17 13/3/15 | 18 22/12/06
19 30/4/09 | 20 28/7/03

Strategy
1 90 | 2 300
3 5000 | 4 160
5 7000 | 6 29 400
7 14 000 | 8 5600
9 390 | 10 23 000
11 6080 | 12 6400
13 109 000 | 14 1420

Space Teacher check drawings
1 rectangle or square
2 trapezium
3 parallelogram or rhombus
4 kite

Problem of the week 1 250 2 840

Unit 9 - Revision

A
1 56 | 2 20 | 3 9 | 4 28
5 54 | 6 9 | 7 7 | 8 2
9 27 | 10 1 | 11 21 | 12 53
13 64 | 14 10 | 15 24 | 16 0
17 35 | 18 45 | 19 8 | 20 7

B
1 5000 | 2 $1.15 | 3 0·7 | 4 F
5 $\frac{4}{8}$ ($\frac{1}{2}$) | 6 1 year | 7 $3.45 | 8 20
9 2·38 | 10 1 | 11 3 510 | 12 63

Answers

13 7·5 cm
14 $\frac{3}{12}$ ($\frac{1}{4}$)
15 $\frac{6}{10}$
16 $2.70
17 $3
18 1, 2, 4, 8, 16
19 T
20 10 days

C
1 49 2 3 3 72 4 20
5 6 months 6 7 7 3:15
8 26 9 37 10 5 11 $6.20
12 44 m 13 150 14 550 cm
15 270° 16 10:50 17 4 18 1st May
19 7:35 pm 20 4 cm

Strategy
1 33 2 71 3 124 4 96
5 66 6 91 7 85 8 29
9 17 10 68 11 132 12 112
13 176 14 1900 15 27 000

Space Teacher check

Time 1

2 11:05

Fractions Teacher check

Number

Factor	7	9	8	9
Factor	9	5	6	6
Multiple	63	45	48	54

Factor	2	11	7	20
Factor	16	8	8	3
Multiple	32	88	56	60

Length
1 50 mm 2 37 mm 3 Teacher check

Unit 10

A
1 21 2 19 3 20 4 20
5 21 6 22 7 22 8 22
9 23 10 20 11 23 12 17
13 24 14 19 15 22 16 19
17 21 18 20 19 17 20 20

B
1 0·01, $\frac{1}{10}$, 0·3, $\frac{1}{2}$, 0·6
2 0·19, $\frac{1}{5}$, $\frac{1}{4}$, 0·77, 0·99
3 $\frac{1}{100}$, 0·10, $\frac{11}{100}$, 0·70, $\frac{100}{100}$
4 1·97, 2·1, 2·3, 3·02
5 $\frac{1}{2}$, $\frac{1}{3}$, $\frac{1}{4}$, $\frac{1}{5}$, $\frac{1}{8}$
6 5·22, 5·2, 2·5, 0·52, 0·25
7 1·8, 1.5, 0·8, 0·45, $\frac{1}{4}$
8 4·1, 1·4, 1·14, $\frac{1}{4}$

C
1 10 2 70 000 3 24 006
4 40 000 + 1000 + 300
5 7000 6 10 7 $2.97 8 1104c
9 840c 10 $30.15 11 < 12 =
13 < 14 = 15 > 16 =
17 = 18 > 19 = 20 >

Strategy
1 150 2 830 3 7150 4 70
5 130 6 155 7 260 8 370
9 550 10 1410

Data Teacher check

Problem of the week
1 3 2 6 3 6 4 15

Unit 11

A
1 29 2 59 3 40 4 43
5 57 6 60 7 33 8 78
9 33 10 75 11 33 12 58
13 69 14 31 15 37 16 68
17 38 18 71 19 47 20 41

B
1 600 2 200 3 900 4 400
5 800 6 200 7 900 8 600
9 300 10 4000 11 2000 12 8000
13 5000 14 1000 15 1000 16 6000
17 3000 18 7000

C
1 $12.45 2 $6.95 3 $10.75 4 $2.30
5 $4.55 6 $11.85 7 $3.35 8 80c
9 $7.45 10 $8.70 11 50c + 10c + 5c
12 20c + 10c + 5c 13 $1 + $1 + 5c
14 $1 + 50c + 50c 15 $2 + $2 + 50c
16 $1 + 10c + 5c 17 10c + 5c + 5c
18 $2 + 50 + 50c 19 $2 + $2 + 5c
20 50c + 20c + 5c

Strategy
1 58 + 60 + 7 = 125
2 94 + 80 + 9 = 183
3 81 + 40 + 7 = 128
4 137 + 60 + 6 = 203
5 91 − 50 − 4 = 37
6 73 − 20 − 8 = 45
7 82 − 60 − 9 = 13
8 120 − 70 − 5 = 45
9 165 − 80 − 7 = 78

Space
1 pentagon 2 triangle 3 hexagon
4 octagon 5 trapezium

Number approximately
A 8140 B 8250 C 8500 D 8800
E 8900

Problem of the week 9:10 pm

Unit 12

A
1 90 2 42 3 20 4 63
5 0 6 54 7 32 8 36
9 21 10 35 11 64 12 60
3 56 14 28 15 16 16 81
17 45 18 0 19 42 20 27

B
1 0·16 2 $\frac{32}{100}$ 3 0·84 4 $\frac{27}{100}$
5 0·56 6 $\frac{13}{100}$ 7 0·7 8 $\frac{8}{100}$
9 $\frac{99}{100}$ 10 0·04 11 0·47 12 0·28
13 $\frac{68}{100}$ 14 $\frac{11}{100}$ 15 0·01 16 $\frac{82}{100}$
17 $\frac{59}{100}$ 18 0·91 19 $\frac{37}{100}$ 20 0·29

C
1 5 yr 2 15 min 3 $2 4 75 yr
5 $10 6 3 7 2 min 8 500 m
9 9 mth 10 $2 11 $6 12 $6.40
13 $15.70 14 $12.10 15 $3.85 16 $15.95
17 $16.10 18 $14.55 19 $11.05 20 $19

Strategy
1 $5\frac{7}{10}$ 2 $9\frac{3}{10}$ 3 $12\frac{8}{10}$ 4 $20\frac{6}{10}$
5 $14\frac{4}{10}$ $(\frac{2}{5})$ 6 $20\frac{2}{10}$ $(\frac{1}{5})$ 7 $18\frac{8}{10}$ $(\frac{4}{5})$ 8 $46\frac{6}{10}$ $(\frac{3}{5})$

Number
1 8 2 12 r 2 3 10 r 4 4 64
5 16 6 9 r 1 7 21 r 1 8 7 r 1

Hundredths 1 0·38, $\frac{38}{100}$ 2 0·09, $\frac{9}{100}$

Problem of the week $30

Unit 13

A
1 9 2 6 3 8 4 5
5 6 6 6 7 1 8 5
9 7 10 3 11 7 12 8
13 9 14 9 15 9 16 3
17 7 18 6 19 4 20 4

B
1 1, 2, 3, 4, 6, 12 2 1, 2, 4, 5, 10, 20
3 1, 3, 5, 15 4 1, 2, 4, 8
5 1, 3, 9, 27
6 1, 2, 3, 4, 6, 9, 12, 18, 36 7 1, 29
8 1, 2, 5, 10, 25, 50 9 14
10 12 11 32 12 54 13 33
14 42 15 64 16 60

C
1 366 2 4 3 30
4 December 5 3 6 75 mins
7 270 8 84 9 85
10 15 11 39 12 160
13 18 14 160 15 71
16 22 17 77 18 17
19 24 20 12

Strategy
1 52 ÷ 2 ÷ 2 = 13 2 88 ÷ 2 ÷ 2 = 22
3 64 ÷ 2 ÷ 2 =16 4 96 ÷ 2 ÷ 2 = 24
5 128 ÷ 2 ÷ 2 = 32 6 260 ÷ 2 ÷ 2 = 65
7 74 ÷ 2 ÷ 2 = $18\frac{1}{2}$ 8 152 ÷ 2 ÷ 2 = 38

Patterns

Day	1	2	3	4	5	6	7	8	9
Pages	23	26	29	32	35	38	41	44	47
Total		49	78	110	145	183	224	268	315

Chance
1 impossible 2 certain
3 impossible 4 possible
5 possible

Problem of the week
1 Marnie's 2 0·96 m

Unit 14

A
1 1·2 2 1·2 3 1 4 0·9
5 1·1 6 1·3 7 1·3 8 0·6
9 0·1 10 0·3 11 0·6 12 0·4
13 1 14 0·7 15 6·3 16 12·7
17 14·3 18 4·3 19 5·6 20 3·6

B
1 21 2 46 3 16 4 42
5 39 6 1 7 1 8 64
9 17 10 16 11 5 12 8
13 15 14 9 15 9 16 $2.50
17 $15 18 $3 19 $12 20 $0.30

C
1 3 mL 2 8 cm 3 29 g
4 512 kg 5 76 kg 6 115 L
7 35 mL 8 7 am 9 612 m
10 9:30 pm 11 545 cm 12 325 mm
13 1160 g 14 8040 m 15 1035 min
16 3·09 km 17 85·4 cm 18 6·43 kg
19 $6\frac{2}{3}$ hrs 20 8·7 m

Strategy
1 160 ÷ 2 ÷ 2 ÷ 2 = 20
2 144 ÷ 2 ÷ 2 ÷ 2 = 18
3 112 ÷ 2 ÷ 2 ÷ 2 = 14
4 192 ÷ 2 ÷ 2 ÷ 2 = 24
5 200 ÷ 2 ÷ 2 ÷ 2 = 25
6 176 ÷ 2 ÷ 2 ÷ 2 = 22

Answers

Volume 1 20 cm^3 2 36 cm^3 3 32 cm^3
Space Teacher check
Problem of the week 24

Unit 15

A
1 10 | 2 3 | 3 42 | 4 6
5 1 | 6 8 | 7 54 | 8 64
9 3 | 10 28 | 11 5 | 12 7
13 2 | 14 56 | 15 7 | 16 0
17 9 | 18 27 | 19 8 | 20 63

B
1 3000 g | 2 4500 g | 3 13 000 g
4 7250 g | 5 1700 g | 6 9550 g
7 5750 g | 8 26 118 g | 9 7 kg
10 12 kg | 11 16 kg | 12 8 kg 500 g
13 10 kg 750 g | 14 43 kg 100 g
15 82 kg 745 g | 16 2 kg 8 g

C
1 06:00 | 2 21:00 | 3 18:00
4 11:00 | 5 22:00 | 6 20:00
7 07:15 | 8 13:30 | 9 17:40
10 09:50 | 11 5:00 am | 12 2 pm
13 11 pm | 14 1 am | 15 10 am
16 4 pm | 17 8:15 pm | 18 7:30 pm
19 6:05 pm | 20 3:10 am

Strategy
1 30 + 23 = 53 | 2 35 + 30 = 65
3 30 + 18 = 48 | 4 20 + 59 = 79
5 20 + 17 = 37 | 6 40 + 46 = 86
7 40 + 36 = 76 | 8 83 + 20 = 103
9 100 + 69 = 169

Space 1 Pyramid 2 Triangular prism
3 Triangular prism
Position north, north-east, east, south-east, south, south-west, west, north-west
Problem of the week 6, 12, 8, 1

Unit 16

A
1 15 | 2 35 | 3 85 | 4 13
5 33 | 6 83 | 7 16 | 8 36
9 86 | 10 12 | 11 32 | 12 72
13 11 | 14 31 | 15 81 | 16 13
17 33 | 18 93 | 19 11 | 20 81

B
1 $\frac{3}{10}$ | 2 $\frac{6}{10}$ | 3 $\frac{8}{10}$ | 4 $\frac{13}{100}$
5 $\frac{72}{100}$ | 6 $\frac{95}{100}$ | 7 $\frac{29}{100}$ | 8 $1\frac{2}{10}$
9 $1\frac{5}{10}$ | 10 $2\frac{28}{100}$ | 11 $\frac{64}{100}$ | 12 $3\frac{73}{100}$
13 $\frac{11}{100}$ | 14 $\frac{36}{100}$ | 15 $5\frac{55}{100}$ | 16 $\frac{99}{100}$
17 $1\frac{1}{100}$ | 18 $\frac{4}{100}$

C
1 decagon | 2 cylinder | 3 scalene
4 angle | 5 rhombus | 6 position
7 obtuse | 8 pyramid | 9 isosceles
10 equilateral | 11 south | 12 west
13 north-east | 14 north-west
15 1·25 L | 16 180° | 17 $375 000
18 2·5 kg | 19 18 mm | 20 91

Strategy Teacher check
Space Teacher check drawings
1 cone 2 triangular pyramid
Problem of the week 4

Unit 17

A
1 60 | 2 190 | 3 2040 | 4 3760
5 160 | 6 280 | 7 560 | 8 2060
9 1380 | 10 4300 | 11 120 | 12 210
13 360 | 14 440 | 15 400 | 16 120
17 420 | 18 720 | 19 350 | 20 900

B
1 4th | 2 12th | 3 8th | 4 2nd
5 9th | 6 12th | 7 5th | 8 21st
9–14 Teacher check

C
1 P = 24 cm A = 32 cm^2
2 P = 48 cm A = 144 cm^2
3 P = 18 cm A = 14 cm^2
4 P = 24 cm A = 27 cm^2
5 cone | 6 triangular prism
7 sphere | 8 triangular pyramid
9 cylinder

Strategy Teacher check
Scale Teacher check
Problem of the week shoes = $100, jacket = $150, shirt = $50

Unit 18

A
1 20 | 2 32 | 3 72 | 4 2·6
5 22 | 6 10 | 7 32 | 8 36
9 86 | 10 20 | 11 6 | 12 160
13 56 | 14 840 | 15 2030 | 16 0
17 0·2 | 18 73 | 19 32 | 20 420

B
1 6000 | 2 200 | 3 2 | 4 6000
5 20 000 | 6 80 | 7 70 000 | 8 200
9 900 000 | 10 7646
11 6000 + 200 + 9 | 12 2000 + 1
13 17 000 | 14 592 000
15 70 000 | 16 280 000

C 1 6000 2 24 007 3 $5.76
4 1303c 5 $17.55 6 $2.95
7 50c + 20c + 5c 8 63
9 75 10 21:40 11 12
12 $8 13 $7.50 14 12 months
15 26 16 35 17 32
18 49 19 60 20 210

Strategy
1 150 2 2960 3 230 4 1030
5 105 6 27 7 $4\frac{7}{10}$ 8 $14\frac{2}{5}$
9 65 10 15 11 45 12 98
13 180 14 630 15 62

Space
1 triangle 2 quadrilateral
3 pentagon 4 hexagon
5 octagon

Number $\frac{12}{100}$, 0·12
Chance Teacher check
Time 18:45
Space Teacher check

Unit 19

A 1 10 2 24 3 6 4 49
5 48 6 35 7 72 8 54
9 21 10 25 11 56 12 40
13 81 14 36 15 0 16 64
17 18 18 27 19 42 20 40

B 1 57 2 99 3 XXXIX 4 LXXII
5 2000 6 300 7 4·89 8 8·5 kg
9 80 10 120 11 6·4 12 7·5
13 92% 14 4% 15 15·7 16 25
17 8 18 44 19 3750 g

C 1 3 2 30 3 52 4 366
5 92 6 72 7 300 8 30
9 10 10 25 11 400 12 1200
13 195 14 48 15 10 16 540
17 19:23 18 2:15 pm
19 09:30 20 8:47 pm

Strategy
1 840 2 970 3 1650 4 3080
5 4600 6 340 7 720 8 1280
9 2060 10 8320 11 540 12 960
13 1150 14 1240 15 3690

Number
1 251 930 2 403 706 3-4 Teacher check

Space Teacher check 1 trapezium 2 kite
Problem of the week Teacher check

Unit 20

A 1 29 2 38 3 29 4 37
5 28 6 31 7 54 8 28
9 27 10 25 11 46 12 57
13 56 14 47 15 43 16 47
17 37 18 63 19 45 20 42

B 1 CC 2 D 3 L 4 CM
5 LXXX 6 CXL 7 LXXVI
8 CCLXXXIII 9 MCC 10 400
11 800 12 20 13 90 14 19
15 165 16 431 17 911 18 999

C 1 8, 8·5 2 1·02, 1·32
3 1·5, 1·75 4 3·9, 4·8
5 7·95, 7·85 6 5·93, 5·63
7 $2.35 8 57c 9 $8.18 10 $1.45
11 $6.01 12 $12.45 13 $6.20 14 $16.55
15 $3.28 16 $14.87

Strategy
1 7 × 8 = 56 2 4 × 10 = 40
3 9 × 11 = 99 4 12 × 5 = 60
5 13 × 10 = 130 6 21 × 5 = 105
7 8 × 20 = 160 8 7 × 100 = 700

Data 1 16, 10 2 64 3 Teacher check
4 Cartoons & sport 5 Teacher check
Problem of the week 15, 7

Unit 21

A 1 6 2 5 3 9 4 11
5 6 6 8 7 8 8 3
9 11 10 7 11 14 12 1
13 0 14 9 15 14 16 22
17 12 18 12 19 19 20 17

B 1 $4\frac{9}{10}$ 2 $5\frac{3}{10}$ 3 $7\frac{2}{10}$ 4 $8\frac{1}{10}$
5 21 6 15 7 36 8 $59\frac{4}{10}$
9 $12\frac{1}{10}$ 10 $5 11 $8 12 $12
13 $1.76 14 $2.49 15 $8.03 16 $9.60
17 $25 18 $13.60

C 1 72 2 495 3 16 4 32
5 240 6 77 7 25 8 7
9 31 10 25 11 11 12 10
13 38 14 11 15 25 16 90
17 11 18 6 19 70 20 100

Answers

Strategy

1. 93 − 60 + 6 = 39
2. 87 − 40 + 1 = 48
3. 163 − 40 + 5 = 128
4. 275 − 60 + 4 = 219
5. 64 − 30 + 2 = 36
6. 81 − 60 + 2 = 23
7. 126 − 50 + 2 = 78
8. 315 − 70 + 3 = 248

Calculator

1. 1521, 1600, 1681, 1764
2. 156, 168, 180, 192
3. 120, 100, 80, 60, 40 or 20

Mass

1 g 2 kg 3 g 4 kg
5 g 6 g

Problem of the week

Brady = 10:50 am Beppi = 10:05 am

Unit 22

A

1 10	2 8	3 8	4 7
5 6	6 8	7 3	8 4
9 1	10 9	11 7	12 6
13 4	14 7	15 5	16 9
17 6	18 8	19 9	20 7

B

1 $\frac{5}{10}$	2 $\frac{2}{3}$	3 $\frac{4}{10}$	4 $\frac{9}{12}$
5 $\frac{3}{4}$	6 $\frac{1}{2}$	7 0·7	8 $\frac{1}{5}$
9 0·9	10 $\frac{1}{10}$	11 $1\frac{3}{4}$	12 7·6
13 $2\frac{1}{2}$	14 37	15 $2\frac{3}{4}$	16 3·3
17 $1\frac{1}{5}$	18 0·91	19 5·3	

C

1 cone	2 rhombus	3 kite
4 angle	5 prism	6 triangle
7 radius	8 isosceles	9 trapezium
10 scalene	11 quadrant	12 sphere
13 obtuse	14 cylinder	15 circle
16 pyramid	17 rectangle	18 diameter
19 parallelogram	20 pentagon	

Strategy

1 27	2 25	3 46	4 37
5 39	6 58	7 58	8 86
9 96	10 57	11 132	12 57

Space Teacher check

Number

1. Colour 9, 18, 27, 36, 45, 54, 63, 72, 81, 90, 99
2. Teacher check
3. The digits always add to 9 except for the last one.

Problem of the week

Suzie gives one to Sam and two to Sean and Sal givestwo to Sean.

Unit 23

A

1 31	2 48	3 65	4 35
5 32	6 25	7 71	8 30
9 7	10 69	11 41	12 36
13 9	14 19	15 51	16 75
17 13	18 55	19 65	20 52

B

1 11	2 6	3 5	4 75
5 88	6 44	7 35	8 6
9 29	10 8·7	11 12·7	12 2·6
13 13·5	14 17·3	15 8·33	16 8·77
17 6·78	18 15·9		

C

1 50c + 20c + 10c 2 $1 + 10c
3 $2 + $1 + 50c + 10c + 5c
4 $5 + 20c + 20c
5 $5 + $2 + $1 + 50c + 20c + 20c + 5c
6 $10 + $5 + 50c
7 $20 + $2 + 50c + 20c + 5c
8 $20 + $10 + 20c + 10c + 5c
9 $100 + $50 + $5 + 50c + 10c
10 $100 + $100 + $10 + $5

11 40	12 38	13 26	14 36
15 22	16 11	17 26	18 15
19 19	20 10		

Strategy

1 $9\frac{8}{10}$	2 $5\frac{3}{10}$	3 $17\frac{4}{10}$	4 $25\frac{1}{10}$
5 $90\frac{7}{10}$	6 $3\frac{81}{100}$	7 $5\frac{74}{100}$	8 $8\frac{16}{100}$
9 $10\frac{28}{100}$	10 $73\frac{5}{100}$	11 $140\frac{2}{10}$	12 $30\frac{6}{100}$
13 90	14 $102\frac{15}{100}$	15 $2206\frac{3}{10}$	

Number 1 $\frac{6}{10}$ 2 $\frac{3}{8}$ 3 $\frac{1}{4}$ 4 $\frac{3}{5}$ 5 $\frac{5}{12}$ 6 $\frac{6}{10}$ and $\frac{3}{5}$

Patterns Teacher check

Problem of the week I am 17.

Unit 24

A
1 8·7 2 9·8 3 11·6 4 17·8
5 13·9 6 20·7 7 6·2 8 6·1
9 14·1 10 4·4 11 9·5 12 16·1
13 12·1 14 8·4 15 11·4 16 16·8
17 18·2 18 18·2 19 18·4 20 20

B Other answers are acceptable.
1 10 2 20 3 10 4 180
5 340 6 150 7 200 8 130
9 200 10 30 11 $7\frac{1}{4}$ 12 $9\frac{4}{9}$
13 $8\frac{2}{3}$ 14 $10\frac{2}{5}$ 15 $9\frac{3}{8}$ 16 $7\frac{2}{7}$
17 $18\frac{1}{2}$ 18 $9\frac{5}{6}$ 19 $4\frac{2}{10}$ 20 $4\frac{9}{20}$

C
1 20 2 5 3 3 4 720
5 11 000 6 8 7 1900 8 3500
9 5·5 10 9 11 4500 12 125
13 6500 14 $4\frac{1}{4}$ 15 185 16 2·25
17 5750 18 7500 19 3500 20 305

Strategy
Other answers are acceptable.
1 320 2 11 3 130 4 400
5 4500 6 200 7 320 8 20
9 6000 10 900 11 140 12 9600
13 2000 14 30 15 350

Space
1 front 2 side 3 top 4 20

Length
1 40 mm, 27 mm, 46 mm, 30 mm, 60 mm
2 203 mm

Problem of the week
A = 2, B = 3, C = 6, D = 8, E = 5,
F = 1, G = 7, H = 4, I = 9

Unit 25

A
1 \$1.10 2 \$2.10 3 \$2310 4 \$2.20
5 \$3.10 6 \$3.12 7 \$3.21 8 3·4
9 3·2 10 1·1 11 1·2 12 1·2
13 1·2 14 1·4 15 1·5 16 1·8
17 2·6 18 3·3 19 4·9 20 8·3

B
1 $\frac{1}{4}$ 2 $\frac{2}{3}$ 3 $\frac{3}{5}$ 4 $\frac{5}{6}$
5 $\frac{3}{10}$ 6 $\frac{1}{5}$ 7 $\frac{6}{10}$ 8 $\frac{7}{8}$
9 $\frac{1}{2}$ 10 $\frac{2}{3}$ 11 $\frac{3}{5}$ 12 $\frac{3}{8}$
13 $\frac{1}{2}$ 14 $\frac{1}{3}$ 15 $\frac{7}{8}$ 16 $\frac{1}{2}$

C
1 acute 2 obtuse 3 right angle
4 reflex 5 obtuse 6 acute
7 obtuse 8 straight 9 reflex
10 revolution 11 50°
12 110° 13 10° 14 70° 15 90°
16 60° 17 100° 18 50°

Strategy
1 121 2 160 3 142 4 183
5 297 6 45 7 36 8 47
9 78 10 239 11 183 12 402
13 45 14 545 15 584

Angles
1 60° 2 25° 3 135° 4 160°
5-7 Teacher check

Calculator 1 D 2 C

Problem of the week
C, D →; ←D; → D, M; ←M; → M, L

Unit 26

A
1 1 2 45 3 4 4 49
5 5 6 48 7 9 8 6
9 7 10 24 11 9 12 9
13 10 14 42 15 32 16 8
17 0 18 63 19 9 20 2

B
1 b 2 d 3 a 4 c
5 h 6 e 7 f 8 g
9 XIV 10 XXXVIII 11 LXV 12 CI
13 72 14 96 15 300

C
1 cm^2 2 m^2 3 m^2 4 ha
5 ha 6 cm^2 7 ha 8 m^2
9 cm^2 10 cm^2 11 22 cm 12 10 m
13 38 mm 14 36 cm
15 38 m 16 62 m
17 46 mm 18 46 cm
19 21·6 m 20 48 cm

Strategy
1 269 2 80 + 40 = 120
3 110 + 66 = 176 4 120 + 96 = 216
5 210 + 147 = 357 6 320 + 128 = 448
7 540 + 108 = 648 8 260 + 130 = 390

Coordinates
1-3 Teacher check
4 Christmas tree

Problem of the week 36

Answers

Unit 27 – Revision

A 1 54 2 34 3 3 4 7
5 34 6 4·2 7 56 8 63
9 39 10 48 11 6 12 17·1
13 1 14 46 15 11·1 16 79
17 0 18 20 19 $1.80 20 48

B 1 2000 2 200 000 3 704
4 69 5 CCLXXXI 6 MD
7 56 8 1, 2, 4, 5, 10, 20 9 11
10 240 11 80c 12 $34.50 13 $12.80
14 $10.20 15 $7\frac{2}{7}$ 16 $\frac{3}{8}$ 17 25
18 3 months 19 20c 20 $2

C 1 240 2 52 3 366 4 1200
5 7 6 $6.85 7 31 8 39
9 37 10 14 11 $2.75 12 $5.30
13 42 14 44 15 81 16 27
17 acute 18 28 m 19 $7.35 20 7·2

Strategy
1 760 2 5070 3 360 4 78
5 39 6 54 7 56 8 $9\frac{3}{10}$
9 $8\frac{21}{100}$ 10 103 11 55 12 309
13 238 14 624 15 140

Space square/rectangular prism

Number 1 $\frac{1}{4}$ 2 $\frac{3}{6}$ ($\frac{1}{2}$) 3 $\frac{3}{8}$ 4 $\frac{1}{3}$

Calculator
1 ($45^2 \rightarrow 54^2$) 2 60, 120, 180 3 27

Space 1 14 2 a b

Unit 28

A 1 0 2 36 3 35 4 54
5 8 6 56 7 16 8 15
9 63 10 32 11 25 12 42
13 36 14 24 15 28 16 40
17 80 18 21 19 81 20 42

B 1 24 2 Feb
3 Teacher Check
4 Teacher Check

C 1 Six thousand 2 Fifty thousand
3 Three tenths 4 Seven
5 Four hundredths 6 Five hundredths
7 Zero (or no tenths)
8 Zero
9 Five thousandths
10 Seven tenths

Strategy
1 292 2 310 + 186 = 496
3 270 + 135 = 405 4 440 + 132 = 572
5 620 + 434 = 1054
6 750 + 150 = 900
7 530 + 424 = 954
8 390 + 117 = 507
9 620 + 558 = 1178

Number
1 $2\frac{1}{4}$ 2 $1\frac{5}{8}$ 3 $3\frac{3}{10}$ 4 $1\frac{3}{5}$ 5 $1\frac{1}{3}$

Space
1 quadrilateral 2 pentagon
3 hexagon 4 heptagon
5 octagon 6 nonagon
7 decagon

Problem of the week 1385

Unit 29

A 1 9 2 6 3 9 4 9
5 6 6 8 7 9 8 5
9 5 10 4 11 7 12 8
13 9 14 8 15 6 16 8
17 9 18 3 19 5 20 7

B 1 99 2 140 3 134 4 109
5 129 6 284 7 123 8 116
9 81 10 273

C 1 odd 2 even/odd 3 even
4 odd 5 even/odd 6 even
7 even 8 odd 9 even 10 odd
11 104 12 1, 2, 4 13 3 14 2
15 90 16 72 17 kg 18 m
19 mL 20 mm

Strategy Other answers are acceptable.
1 50 2 180 3 90 4 25
5 240 6 110 7 220 8 6
9 800 10 200 11 220 12 60 000

Space
1 reflex, 270° 2 straight, 180°
3 acute, 27° 4 obtuse, 111°

Time
1 3:23 pm 2 7:02 am 3 8:51 pm
4 11:19 am 5 4:48 pm

Problem of the week
1 square 2 heart, dot, star and arrow
3 dot 4 heart

Unit 30

A
1 27 2 25 3 15 4 33
5 67 6 46 7 62 8 15
9 56 10 49 11 55 12 5
13 53 14 55 15 35 16 57
17 39 18 83 19 27 20 50

B
1 100 2 130 3 130 4 120
5 300 6 60 7 20 8 30
9 4, 3·9 10 9, 8·6 11 16, 16·2
12 12, 11·9 13 40, 38·1 14 7, 7·2
15 8, 8·2 16 7, 7·3

C
1 24°C 2 1°C 3 5°C 4 10°C
5 2°C 6-10 Teacher check
11 twelve million
12 three million
13 one hundred thousand
14 fifteen million and five
15 one million one hundred
16 2 005 000
17 8 000 010
18 500 000
19 14 000 000
20 1 000 100

Strategy
1 17, 170, 1700 2 12, 120, 1200
3 21, 210, 2100 4 24, 240, 2100
5 26, 260, 2600 6 14, 1400, 14 000

Measurement
1 12 2 16 3 20 4 24

Position
1 north-west 2 south
3 south-west 4 south-east
5 north-east

Problem of the week

Unit 31

A
1 29 2 31 3 28 4 27
5 38 6 27 7 26 8 37
9 33 10 39 11 34 12 32
13 39 14 35 15 48 16 43
17 51 18 39 19 43 20 117

B
1 $3.50 2 70c 3 $4.80 4 $1.50
5 $9.20 6 $34 7 $18.60 8 $52.90
9 $80.30 10 $116.70 11 $9.80
12 $11.20 13 $3.00 14 $5
15 $10.50 16 $24.90 17 $24
18 $30 19 $22 20 $69

C
1 5 2 2 3 3
4 7 5 6 6-15 Teacher check

Strategy
1 85 2 91 3 133 4 103
5 101 6 120 7 119 8 117
9 92 10 94 11 74 12 71
13 106 14 91 15 93

Number
1 $370 2 $555 3 $666 4 $444
5 $185 6 $651.20

Money 1 b 2 a 3 b

Problem of the week
∠A = 40° ∠B = 80° ∠C = 60°

Unit 32

A
1 21 2 33 3 55 4 70
5 47 6 86 7 46 8 35
9 28 10 67 11 87 12 79
13 70 14 27 15 85 16 83
17 73 18 71 19 82 20 61

B
1 17 2 9 3 11 4 32
5 2 6 17 7 8 8 35
9 1 10 27 11 4 12 22
13 30 14 41 15 25 16 140
17 17 18 32 19 2 20 16

C
1 d 2 f 3 a 4 b
5 h 6 c 7 e 8 j
9 g 10 i 11 30 cm, 54 cm²
12 32 m, 63 m² 13 40 mm, 99 mm²
14 56 km, 160 km² 15 48 cm, 119 cm²

Strategy
1 32 2 26 3 48 4 63
5 88 6 52 7 60 8 89
9 37 10 111 11 259 12 163
13 92 14 126 15 34

Number 0·4, 0·5, 0·7, 0·8, 1·1, 1·3, 1·4, 1·6

Data Teacher check

Problem of the week 1 45 × 8 = 360
2 19 × 62 = 1178

Answers

Unit 33

A 1 2·5 2 3·8 3 6·7 4 9·9
5 16·1 6 12·7 7 8·4 8 2·3
9 5·3 10 8 11 4·8 12 1·7
13 4·2 14 0·7 15 3·5 16 4
17 4·9 18 0·8 19 10·9 20 5·6

B 1 $24 2 $81 3 $1.65 4 $3.40
5 $1.86 6 $7.60 7 $8.40 8 $27
9 $22.80 10 $73.35 11 5 12 3
13 4 14 49 15 30 16 9
17 50 18 19

C 1 mL 2 kg 3 cm 4 min
5 mm 6 L 7 mL 8 km
9 g 10 m 11-20 Teacher check

Strategy
1 13 2 24 3 33 4 54
5 12 6 15 7 21 8 35
9 44 10 19 11 205 12 255

Measurement 1 28 m² 2 31 cm² 3 51 m²

Position
1 A 2 H 3 B 4 G
5 F 6 B 7 B 8 H
9 H 10 A

Problem of the week 4

Unit 34

A 1 16 2 7 3 8 4 16
5 56 6 100 7 7 8 24
9 9 10 9 11 42 12 9
13 9 14 8 15 3 16 54
17 24 18 9 19 35 20 7

B 1 115 2 152 3 150 4 $52\frac{1}{2}$
5 400 6 187 7 47 8 314
9 401 10 137 11 $2.50 12 $8.70
13 $2.05 14 $14.40 15 $39.20 16 $9.45
17 55c 18 $1.15 19 $43.90 20 45c

C 1 cylinder
2 triangular prism
3 square/rectangular pyramid
4 cone 5 cube 6 32 7 750
8 18 9 3 10 180 11 135
12 1750 13 330 14 1950 15 91

Strategy
1 300 2 2000 ÷ 4 = 500
3 2400 ÷ 4 = 600 4 4400 ÷ 4 = 1100
5 3600 ÷ 4 = 900 6 2800 ÷ 4 = 700
7 1400 ÷ 4 = 350 8 2100 ÷ 4 = 525
9 2300 ÷ 4 = 575

Space 1 324° 2 reflex 3 Teacher check

Position
1 (0, 4) 2 (1, 3) 3 (3, 1)
4 (5, 2) 5 (5, 0)

Space 5 diagonals

Problem of the week 267

Unit 35 – Revision

A 1 42 2 39 3 45 4 87
5 1 6 34 7 25 8 36
9 58 10 73 11 72 12 4·2
13 8 14 6·4 15 15·3 16 8
17 70 18 19 19 2 20 63

B 1 3 2 30 000 3 $\frac{4}{8}$ 4 $\frac{3}{4}$
5 north 6 20th 7 3rd 8 12th
9 6 10 1·9 11 $2\frac{7}{8}$
12 north-west 13-14 Teacher check
15 $10.53 16 $41\frac{9}{10}$ 17 63
18 3

C 1 44·05 2 8 r 1 3 180° 4 5
5 $1\frac{1}{4}$ 6 cube 7 $6.85 8 $28.60
9 210 10 180 11 19:00
12 $475 000 13 73
14 92 15 2°C 16 possible
17 12:45 18 38 cm, 78 cm²
19 Teacher check 20 225

Strategy
1 34 2 151 3 23, 230, 2300
4 800 5 200 6 70 7 184
8 191 9 1260 10 37 11 36
12 400 13 600

Space
1 acute, 41° 2 reflex, 250°
3 obtuse, 130°

Measurement 1 7:09 am, 10:05 pm
2 17:00, 18:30

Measurement 1 40 m 2 80 m²

Number 1 $260 2 $416
3 $390 4 $364

Strategy
Associative law

Two or more numbers can be multiplied in any order.
eg $7 \times 2 \times 5 = 7 \times 5 \times 2$ or $2 \times 5 \times 7$ or $5 \times 2 \times 7$ or $2 \times 7 \times 5$

Write two different ways to multiply each one.

1 $9 \times 3 \times 8$ = ______________ or ______________

2 $4 \times 5 \times 2$ = ______________ or ______________

3 $6 \times 8 \times 5$ = ______________ or ______________

4 $7 \times 6 \times 9$ = ______________ or ______________

5 $15 \times 17 \times 2$ = ______________ or ______________

6 $12 \times 8 \times 4$ = ______________ or ______________

7 $99 \times 6 \times 10$ = ______________ or ______________

8 $15 \times 18 \times 20$ = ______________ or ______________

Score

Scale

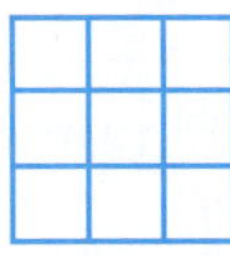

1:2

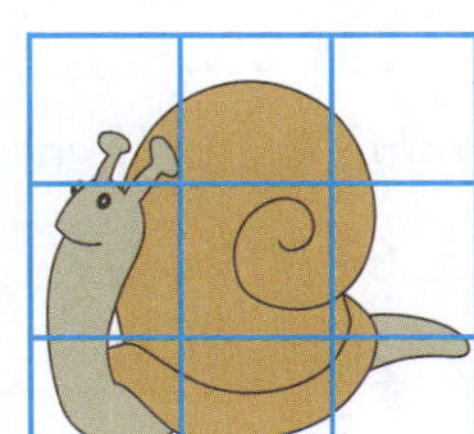

2:1

Enlarge and reduce the drawing of the snail to scale.

Problem of the week

I went shopping. The shoes I bought cost twice as much as the shirt which cost one third as much as the jacket. The scarf was only $15. Altogether I spent $315.

What was the cost of:

the shoes? __________ the shirt? __________ the jacket? __________

Unit 18 Revision

A

1 7 + 8 + 6 = ______
2 6 × 4 + 8 = ______
3 9 × 8 = ______
4 4·1 − 1·5 = ______
5 8 + 9 + 5 = ______
6 100 ÷ 10 = ______
7 3 × 7 + 11 = ______
8 17 + 19 = ______
9 37 + 49 = ______
10 9 + 9 + 2 = ______
11 42 ÷ 7 = ______
12 8 × 20 = ______
13 7 × 8 = ______
14 42 × 20 = ______
15 203 × 10 = ______
16 6 × 0 = ______
17 0·9 − 0·7 = ______
18 8 × 8 + 9 = ______
19 7 + 12 + 13 = ______
20 7 × 60 = ______

Score

B Value of the underlined numeral.

1 3$\underline{6}$ 105 ______
2 90 $\underline{2}$76 ______
3 534$\underline{2}$ ______
4 11$\underline{6}$ 850 ______
5 $\underline{2}$01 478 ______
6 693 0$\underline{8}$4 ______
7 $\underline{7}$5 516 ______
8 6$\underline{2}$09 ______
9 $\underline{9}$7 741 ______
10 7000 + 600 + 40 + 6 = ______
11 26 209 = 20 000 + ______ + ______ + ______
12 302 511 = 300 000 + ______ + 500 + 10 + ______

Round to the nearest thousand.

13 17 298 ______
14 591 802 ______

Round to the nearest ten thousand.

15 65 503 ______
16 283 969 ______

Score

C

1 value of 6 in 46 015 ______
2 20 000 + 4000 + 7 = ______
3 576c = $ ______
4 $13.03 = ______ c
5 $2.45. Change from $20 ______
6 $17.05. Change from $20 ______
7 3 coins to make 75c ______
8 days in 9 weeks ______
9 seconds in $1\frac{1}{4}$ mins ______
10 9:40 pm in 24-hr time ______
11 50% of 2 dozen = ______
12 $3.05 + $4.95 = ______
13 75% of $10 = ______
14 100% of 1 year = ______
15 total 6, 11 and 9 ______
16 subtract 26 from 61 ______
17 Product of 8 and 4 ______
18 next multiple of 7 after 42 ______
19 product of 4 and 15 ______
20 product of 7 and 30 ______

Score

Targeting Mental Maths Year 5 • 978-1-922887-27-6

Strategy

Use the strategies you have learnt.

1 15 × 10 = ________
2 296 × 10 = ________
3 46 × 5 = ________
4 206 × 5 = ________
5 57 + 48 = ________
6 83 – 56 = ________
7 47 ÷ 10 = ________
8 72 ÷ 5 = ________
9 260 ÷ 4 = ________
10 120 ÷ 8 = ________
11 19 + 15 + 11 = ________
12 23 + 48 + 27 = ________
13 15 × 6 × 2 = ________
14 7 × 9 × 10 = ________
15 310 ÷ 5 = ________

Score

Space

Name these shapes.

1 
2 
3 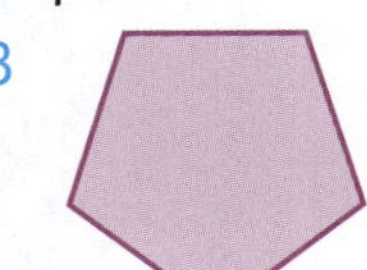
4 
5 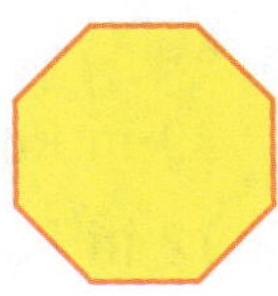

________ ________ ________ ________ ________

Number

Write the coloured part as a fraction and as a decimal.

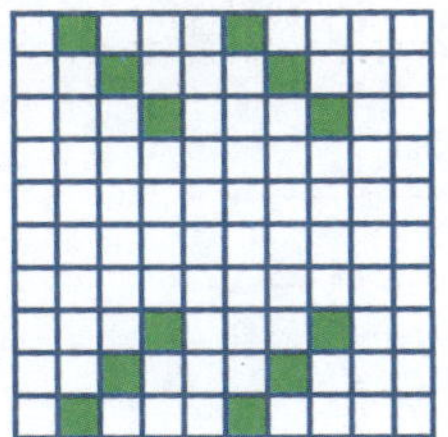

________ ________

Chance

1 Write something that has a fifty-fifty chance of happening.

____

____

2 Write something that has a very slight chance of happening.

____

____

Time

What 24-hour time is half an hour later than 6:15 pm?

Space

I am a quadrilateral with equal sides. My perimeter is 28 cm. How long is each of my sides?

____

Unit 19

A

1. 2 × 5 = ______
2. 4 × 6 = ______
3. 6 × 1 = ______
4. 7 × 7 = ______
5. 6 × 8 = ______
6. 5 × 7 = ______
7. 9 × 8 = ______
8. 6 × 9 = ______
9. 3 × 7 = ______
10. 5 × 5 = ______
11. 7 × 8 = ______
12. 4 × 10 = ______
13. 9 × 9 = ______
14. 6 × 6 = ______
15. 6 × 0 = ______
16. 8 × 8 = ______
17. 6 × 3 = ______
18. 3 × 9 = ______
19. 6 × 7 = ______
20. 5 × 8 = ______

Score

B

1. LVII = ______
2. XCIX = ______
3. 39 in Roman numerals ______
4. 72 in Roman numerals ______
5. 2498 to the nearest thousand ______
6. 309 to the nearest hundred ______
7. 1·64 + 3·25 = ______
8. 8500 g = ______ kg
9. estimate 43 + 39 ______
10. estimate 167 – 52 ______
11. 9 – 2·6 = ______
12. 4 + 3·5 = ______
13. 0·92 = ______ %
14. $\frac{4}{100}$ = ______ %
15. 7·3 + 8·4 = ______
16. ______ + 17 = 42
17. 15 × ______ = 120
18. 78 – ______ = 34
19. $3\frac{3}{4}$ kg = ______ g

Score

C How many?

1. months in Spring ______
2. days in November ______
3. weeks in a year ______
4. days in a leap year ______
5. days in winter ______
6. hours in 3 days ______
7. seconds in 5 minutes ______
8. days in 6 school weeks ______
9. weeks in 5 fortnights ______
10. years in $2\frac{1}{2}$ decades ______
11. years in 4 centuries ______
12. months in 1 century ______
13. minutes in $3\frac{1}{4}$ hours ______
14. minutes between 1:14 and 2:02 ______
15. hours between 9:00 am and 7 pm ______
16. seconds between 11:41 am and 11:50 am ______
17. write 7:23 pm in 24-hour time ______
18. write 14:15 in 12-hour time ______
19. write 9:30 am in 24-hour time ______
20. write 20:47 in 12-hour time ______

Score

Unit 19

Strategy
×10, ×20

× by 10 – add a 0 eg 703 × 10 = 7030
× by 20 – double and add 0 eg 58 × 20 = (58 × 2)0
= 1160

1 84 × 10 = ________
2 97 × 10 = ________
3 165 × 10 = ________
4 308 × 10 = ________
5 460 × 10 = ________
6 17 × 20 = ________
7 36 × 20 = ________
8 64 × 20 = ________
9 103 × 20 = ________
10 416 × 20 = ________
11 18 × 30 = ________
12 24 × 40 = ________
13 23 × 50 = ________
14 31 × 40 = ________
15 123 × 30 = ________

Score

Number

What are these numbers? Draw these numbers.

1

2
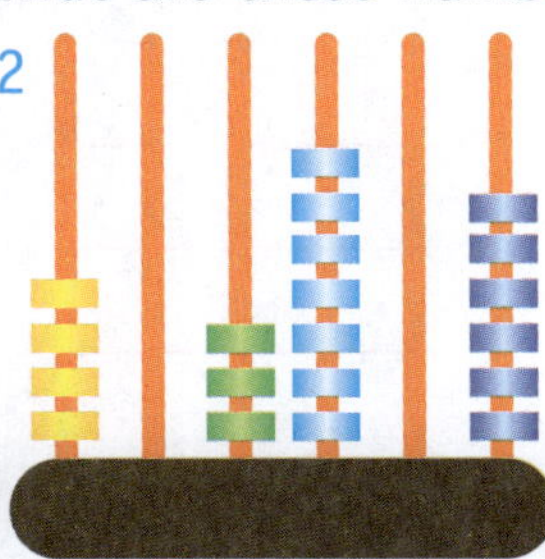

3

290 165

4

145 003

Space

Draw me.

1 I have 4 sides. One pair are parallel.

2 I am a quadrilateral with 2 pairs of equal adjacent sides.

Problem of the week

Choose 4 digits, eg 3 5 1 9

Make the largest number you can.

Make the smallest number you can.

Find the difference.

Use the digits to make the largest number and the smallest number.

Find the difference.

Continue the pattern.

When does it end? ________

Try other 4-digit numbers.

$$\begin{array}{r} 9531 \\ -\,1359 \\ \hline 8172 \\ \hline \end{array}$$

$$\begin{array}{r} 8721 \\ -\,1278 \\ \hline 7443 \\ \hline \end{array}$$

Unit 20

A

1. 7 + 9 + 13 = ______
2. 18 + 12 + 8 = ______
3. 16 + 9 + 4 = ______
4. 17 + 11 + 9 = ______
5. 15 + 8 + 5 = ______
6. 11 + 17 + 3 = ______
7. 9 + 24 + 21 = ______
8. 8 + 15 + 5 = ______
9. 14 + 7 + 6 = ______
10. 7 + 13 + 5 = ______
11. 26 + 8 + 12 = ______
12. 10 + 27 + 20 = ______
13. 31 + 16 + 9 = ______
14. 6 + 17 + 24 = ______
15. 21 + 9 + 13 = ______
16. 17 + 18 + 12 = ______
17. 14 + 7 + 16 = ______
18. 9 + 13 + 41 = ______
19. 17 + 13 + 15 = ______
20. 8 + 22 + 12 = ______

Score

B

Change to Roman numerals.

1. 200 ______
2. 500 ______
3. 50 ______
4. 900 ______
5. 80 ______
6. 140 ______
7. 76 ______
8. 283 ______
9. 1200 ______

Change to Hindu-Arabic numerals.

10. CD ______
11. DCCC ______
12. XX ______
13. XC ______
14. XIX ______
15. CLXV ______
16. CDXXXI ______
17. CMXI ______
18. CMXCIX ______

Score

C

Complete these decimal counting sequences.

1. 6·5, 7, 7·5 ______, ______
2. 0·12, 0·42, 0·72, ______, ______
3. 0·75, 1, 1·25, ______, ______
4. 3·6, ______, 4·2, 4·5, ______
5. 8·25, 8·15, 8·05, ______, ______
6. 6·03, ______, 5·83, 5·73, ______

7. \$7.65 + ______ = \$10
8. \$9.43 + ______ = \$10
9. \$1.82 + ______ = \$10
10. \$8.55 + ______ = \$10
11. \$3.99 + ______ = \$10
12. \$7.55 + ______ = \$20
13. \$13.80 + ______ = \$20
14. \$3.45 + ______ = \$20
15. \$16.72 + ______ = \$20
16. \$5.13 + ______ = \$20

Score

Strategy

Distributive law

eg $3 \times 8 + 3 \times 4$ (= $3 \times (8 + 4)$) = 3×12
= 36

1 $7 \times 6 + 7 \times 2$ = 7 x 8 ______
= ______

2 $4 \times 7 + 4 \times 3$ = ______
= ______

3 $9 \times 5 + 9 \times 6$ = ______
= ______

4 $12 \times 2 + 12 \times 3$ = ______
= ______

5 $13 \times 4 + 13 \times 6$ = ______
= ______

6 $21 \times 3 + 21 \times 2$ = ______
= ______

7 $8 \times 11 + 8 \times 9$ = ______
= ______

8 $7 \times 44 + 7 \times 56$ = ______
= ______

Score

Data

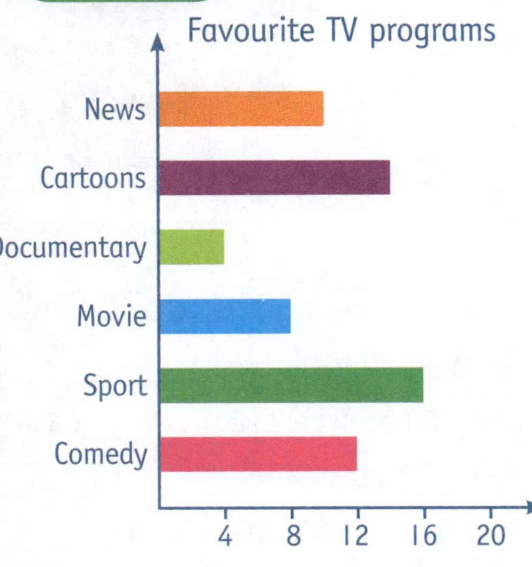

1 How many people preferred to watch Sport? ______ News? ______

2 How many people were surveyed? ______

3 Who would find this survey useful? ______

______

4 If you were advertising goods during which programs would you advertise?

______

5 Is this a good survey? ______
Why? ______

______

Problem of the week

16 people are in a singles tennis tournament. When a player loses they are out of the tournament. How many games are played to find the winner? ______ (Hint: Draw a diagram.)

If they were playing doubles how many games would be played? ______

Unit 21

A

1. 19 − 6 − 7 = ______
2. 22 − 8 − 9 = ______
3. 17 − 3 − 5 = ______
4. 25 − 8 − 6 = ______
5. 18 − 9 − 3 = ______
6. 23 − 7 − 8 = ______
7. 19 − 4 − 7 = ______
8. 26 − 12 − 11 = ______
9. 24 − 8 − 5 = ______
10. 25 − 5 − 13 = ______
11. 32 − 9 − 9 = ______
12. 15 − 6 − 8 = ______
13. 29 − 10 − 19 = ______
14. 18 − 5 − 4 = ______
15. 31 − 9 − 8 = ______
16. 34 − 7 − 5 = ______
17. 28 − 5 − 11 = ______
18. 22 − 6 − 4 = ______
19. 33 − 4 − 10 = ______
20. 37 − 8 − 12 = ______

Score

B

1. 49 ÷ 10 = ______
2. 53 ÷ 10 = ______
3. 72 ÷ 10 = ______
4. 81 ÷ 10 = ______
5. 210 ÷ 10 = ______
6. 150 ÷ 10 = ______
7. 360 ÷ 10 = ______
8. 594 ÷ 10 = ______
9. 121 ÷ 10 = ______

Change to dollars.

10. 500c ______
11. 800c ______
12. 1200c ______
13. 176c ______
14. 249c ______
15. 803c ______

16. 960c ______

17. 2500c ______

18. 1360c ______

Score

C

1. product of 9 and 8 ______
2. sum of 15 and 33 ______
3. difference between 18 and 34 ______
4. total 11, 7 and 14 ______
5. ten times twenty-four ______
6. 7 lots of 11 ______
7. subtract 17 from 42 ______
8. divide 49 by 7 ______
9. sum of 12 and 19 ______
10. take 18 from 43 ______
11. how many nines in 45 ______
12. share 120 between 12 ______ each
13. 3 lots of 8 plus 14 ______
14. add 7 to 15 and $\frac{1}{2}$ the answer ______
15. 53 less 28 ______
16. product of 5, 9 and 2 ______
17. 28 minus the sum of 9 and 8 ______
18. 18 less than half of 48 ______
19. a quarter of 40 times 7 ______
20. square 10 ______

Score

Targeting Mental Maths Year 5 • 978-1-922887-27-6

Unit 21

Strategy
Compensation strategy

eg 138 + 27 = 138 + 30 − 3
= 165

249 − 68 = 249 − 70 + 2
= 181

1 93 − 54 = ______
= ______

2 87 − 39 = ______
= ______

3 163 − 35 = ______
= ______

4 275 − 56 = ______
= ______

5 64 − 28 = ______
= ______

6 81 − 58 = ______
= ______

7 126 − 48 = ______
= ______

8 315 − 67 = ______
= ______

Score

Calculator

1 Find all the square numbers between 1500 and 1800.

______

______

2 Find five multiples of 12 between 150 and 200.

______

3 Find five multiples of 4 which are also multiples of 5 less than 125.

______

Mass

Would you use gram or kilogram to find the mass?

1  ______

2  ______

3  ______

4  ______

5  ______

6  ______

Problem of the week

Brady and Beppi live 18 km from the beach. Brady rides her bike at 12 km per hour and Beppi rides his at 8 km per hour. They want to arrive at the beach together at 12:20 pm. When should each leave home?

Brady ______ Beppi ______

Unit 22

A

1 100 ÷ 10 = ______
2 64 ÷ 8 = ______
3 56 ÷ 7 = ______
4 42 ÷ 6 = ______
5 54 ÷ 9 = ______
6 48 ÷ 6 = ______
7 27 ÷ 9 = ______
8 36 ÷ 9 = ______
9 7 ÷ 7 = ______
10 81 ÷ 9 = ______
11 56 ÷ 8 = ______
12 42 ÷ 7 = ______
13 32 ÷ 8 = ______
14 63 ÷ 9 = ______
15 45 ÷ 9 = ______
16 63 ÷ 7 = ______
17 48 ÷ 8 = ______
18 72 ÷ 9 = ______
19 54 ÷ 6 = ______
20 49 ÷ 7 = ______

Score

B

1 $\frac{1}{2} = \frac{\square}{10}$
2 $\frac{\square}{3} = \frac{4}{6}$
3 $\frac{2}{5} = \frac{4}{\square}$
4 $\frac{3}{4} = \frac{\square}{12}$
5 $\frac{6}{8} = \frac{\square}{4}$
6 0·5 = $\frac{\square}{2}$
7 $\frac{7}{10}$ = 0·____
8 0·2 = $\frac{1}{\square}$
9 $\frac{9}{10}$ = 0·____
10 0·1 = $\frac{\square}{10}$
11 1, $1\frac{1}{4}$, $1\frac{1}{2}$, ______ , 2
12 10, 9·2, 8·4, ______ , 7
13 3, ______ , 2, $1\frac{1}{2}$, 1
14 40, 38·5, ______ , 35·5, 34
15 $\frac{3}{4}$, $1\frac{1}{4}$, $1\frac{3}{4}$, $2\frac{1}{4}$, ______
16 0·5, 1·2, 1·9, 2·6, ______
17 2, $1\frac{3}{5}$, ______ , $\frac{4}{5}$, $\frac{2}{5}$
18 1, 0·97, 0·94, ______ , 0·88
19 3·1, 4·2, ______ , 6·4

Score

C Unscramble the letters to find a 2D or 3D word.

1 eocn ______
2 hbuorms ______
3 eitk ______
4 galen ______
5 mspir ______
6 gaintler ______
7 saduri ______
8 ceelisoss ______
9 ziterupma ______
10 nesclae ______
11 raqduatn ______
12 hesrep ______
13 suetbo ______
14 ylniecdr ______
15 reclic ______
16 rapimdy ______
17 clangreet ______
18 midateer ______
19 lllrraaaepogm ______
20 pegtanno ______

Score

Unit 22

Strategy
Look for tens

eg 8 + 13 + 2 = 23 | 31 + 24 + 16 = 71

1 9 + 7 + 11 = ________
2 15 + 8 + 2 = ________
3 21 + 9 + 16 = ________
4 16 + 17 + 4 = ________
5 13 + 19 + 7 = ________
6 18 + 15 + 25 = ________
7 21 + 18 + 19 = ________
8 7 + 43 + 36 = ________
9 35 + 46 + 15 = ________
10 27 + 16 + 14 = ________
11 93 + 32 + 7 = ________
12 17 + 18 + 22 = ________

Score

Space
Draw the nets.

1
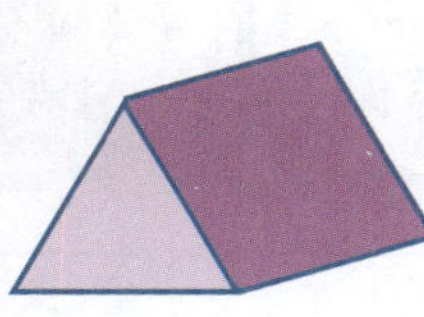

2
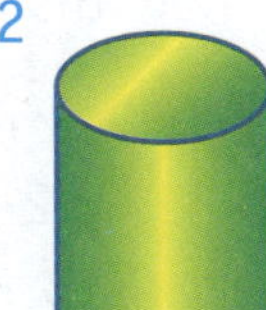

Number
1 Colour all the multiples of 9 in yellow.
2 Write about the patterns you can see.

____

____

____

3 Add the digits of the numbers that you coloured yellow. What do you notice?

____

____

1	2	3	4	5	6	7	8	9	10
11	12	13	14	15	16	17	18	19	20
21	22	23	24	25	26	27	28	29	30
31	32	33	34	35	36	37	38	39	40
41	42	43	44	45	46	47	48	49	50
51	52	53	54	55	56	57	58	59	60
61	62	63	64	65	66	67	68	69	70
71	72	73	74	75	76	77	78	79	80
81	82	83	84	85	86	87	88	89	90
91	92	93	94	95	96	97	98	99	100

Problem of the week
Sam, Susie, Sal and Sean all have sweets. They want to have equal numbers. Sam has 16, Susie has 20, Sal has 19 and Sean has 13.

Work out who gives sweets to whom!

Unit 23

A

1. 3 × 8 + 7 = ______
2. 6 × 9 − 6 = ______
3. 7 × 8 + 9 = ______
4. 10 × 4 − 5 = ______
5. 7 × 4 + 4 = ______
6. 4 × 3 + 13 = ______
7. 7 × 9 + 8 = ______
8. 7 × 6 − 12 = ______
9. 4 × 4 − 9 = ______
10. 8 × 8 + 5 = ______
11. 8 × 6 − 7 = ______
12. 3 × 7 + 15 = ______
13. 9 × 0 + 9 = ______
14. 6 × 4 − 5 = ______
15. 5 × 8 + 11 = ______
16. 9 × 9 − 6 = ______
17. 2 × 9 − 5 = ______
18. 7 × 7 + 6 = ______
19. 8 × 9 − 7 = ______
20. 6 × 7 + 10 = ______

Score

B

1. ______ + 9 = 20
2. 72 ÷ ______ = 12
3. ______ × 20 = 100
4. 205 − ______ = 130
5. ______ ÷ 8 = 11
6. ______ + 25 = 69
7. 81 − ______ = 46
8. 15 × ______ = 90
9. ______ + 52 = 81
10. 6·4 + 2·3 = ______
11. 8·9 + 3·8 = ______
12. 1·7 + 0·9 = ______
13. 5·2 + 8·3 = ______
14. 12·6 + 4·7 = ______
15. 3·02 + 5·31 = ______
16. 1·64 + 7·13 = ______
17. 5·26 + 1·52 = ______
18. 8·79 + 7·11 = ______

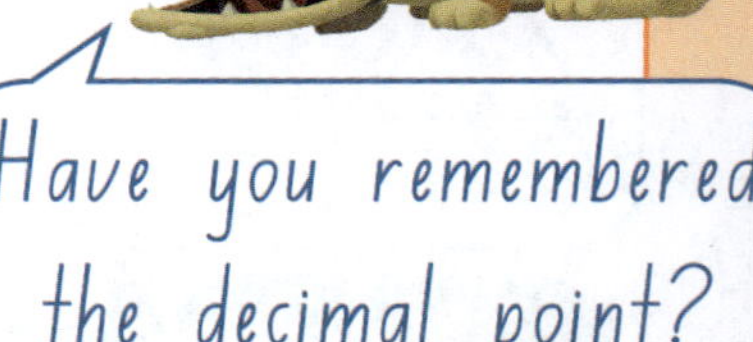

Score

C Write the least number of notes and/or coins to make:

1. 80c ______
2. $1.10 ______
3. $3.65 ______
4. $5.40 ______
5. $8.95 ______
6. $15.50 ______
7. $22.75 ______
8. $30.35 ______
9. $155.60 ______
10. $215.00 ______

How many:

11. 5c coins in $2? ______
12. 10c coins in $3.80? ______
13. 20c coins in $5.20? ______
14. 50c coins in $18? ______
15. $2 coins in $44? ______
16. $5 notes in $55? ______
17. $10 notes in $260? ______
18. $20 notes in $300? ______
19. $50 notes in $950? ______
20. $100 notes in $1000? ______

Score

Strategy
÷ by 10
÷ by 100

To divide by 10 the last digit is the remainder.
eg $76 \div 10 = 7\frac{6}{10}$

To divide by 100 the last two digits are the remainder.
eg $1364 \div 100 = 13\frac{64}{100}$

1 98 ÷ 10 = ______
2 53 ÷ 10 = ______
3 174 ÷ 10 = ______
4 251 ÷ 10 = ______
5 907 ÷ 10 = ______
6 381 ÷ 100 = ______
7 574 ÷ 100 = ______
8 816 ÷ 100 = ______
9 1028 ÷ 100 = ______
10 7305 ÷ 100 = ______
11 1402 ÷ 10 = ______
12 3006 ÷ 100 = ______
13 900 ÷ 10 = ______
14 10 215 ÷ 100 = ______
15 22 063 ÷ 10 = ______

Score

Number

Name the fraction coloured.

1 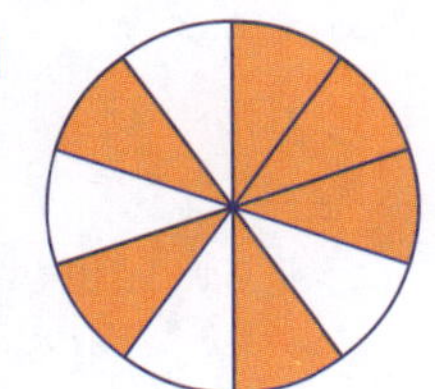 ______

2 ______

3 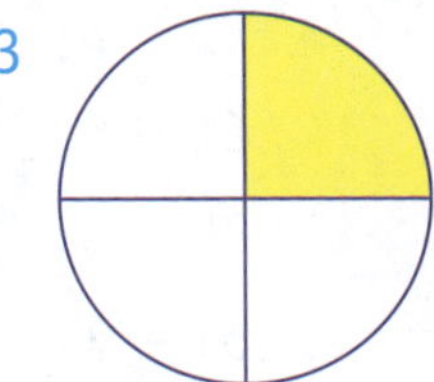 ______

4 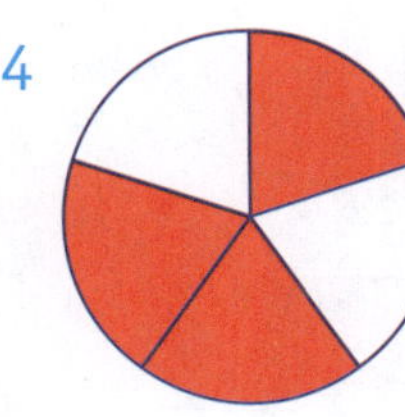 ______

5 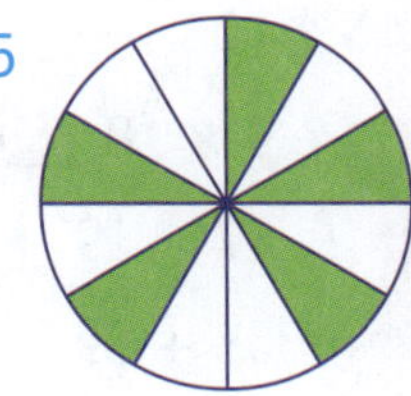 ______

6 Which fractions are equivalent? ______

Patterns

Using 12 ,
8 ,
6  and
4  beads
make a pattern on the string.

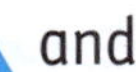

Problem of the week

What am I? I am not a multiple of 2 or 3. I am a teen number.
My digits do not add to 10. The difference between my digits is more than 2.
I am ______.

Unit 24

A

1 6·3 + 2·4 = ________
2 8·3 + 1·5 = ________
3 7·2 + 4·4 = ________
4 9·7 + 8·1 = ________
5 10·5 + 3·4 = ________
6 12·4 + 8·3 = ________
7 1·5 + 4·7 = ________
8 3·5 + 2·6 = ________
9 8·4 + 5·7 = ________
10 1·8 + 2·6 = ________
11 5·6 + 3·9 = ________
12 10·2 + 5·9 = ________
13 7·3 + 4·8 = ________
14 6·9 + 1·5 = ________
15 4·5 + 6·9 = ________
16 9·1 + 7·7 = ________
17 8·8 + 9·4 = ________
18 11·3 + 6·9 = ________
19 12·7 + 5·7 = ________
20 9·6 + 10·4 = ________

Score

B

Estimate only.

1 83 ÷ 9 ________
2 78 ÷ 4 ________
3 103 ÷ 9 ________
4 9 × 18 ________
5 17 × 19 ________
6 63 + 89 ________
7 88 + 112 ________
8 203 – 77 ________
9 364 – 159 ________
10 212 ÷ 7 ________

Write the remainder as a fraction.

11 29 ÷ 4 ________
12 85 ÷ 9 ________
13 26 ÷ 3 ________
14 52 ÷ 5 ________
15 75 ÷ 8 ________
16 51 ÷ 7 ________
17 37 ÷ 2 ________
18 59 ÷ 6 ________
19 42 ÷ 10 ________
20 89 ÷ 20 ________

Score

C

1 200 mm = ________ cm
2 500 cm = ________ m
3 3000 g = ________ kg
4 72 cm = ________ mm
5 11 kg = ________ g
6 8000 mL = ________ L
7 19 m = ________ cm
8 $3\frac{1}{2}$ L = ________ mL
9 55 mm = ________ cm
10 9000 m = ________ km
11 $4\frac{1}{2}$ km = ________ m
12 $1\frac{1}{4}$ m = ________ cm
13 $6\frac{1}{2}$ L = ________ mL
14 4250 g = ________ kg
15 $18\frac{1}{2}$ cm = ________ mm
16 2250 m = ________ km
17 $5\frac{3}{4}$ km = ________ m
18 7·5 kg = ________ g
19 3·5 L = ________ mL
20 30·5 cm = ________ mm

Score

Unit 24

Strategy

To estimate – round to tens or hundreds

eg $215 \div 9 \rightarrow 200 \div 10 = 20$ Estimate = 20

$37 \times 22 \rightarrow 40 \times 20 = 800$ Estimate = 800

$97 + 46 + 384 \rightarrow 100 + 50 + 400 = 550$ Estimate = 550

Estimate only.

1 39×8 ______

2 $115 \div 9$ ______

3 $214 - 73$ ______

4 $368 + 59$ ______

5 456×12 ______

6 $69 + 138$ ______

7 $523 - 174$ ______

8 $381 \div 19$ ______

9 203×28 ______

10 $528 + 369$ ______

11 $76 + 15 + 43 =$ ______

12 $961 \times 11 =$ ______

13 $18 \times 9 \times 11 =$ ______

14 $857 \div 29 =$ ______

15 $615 - 186 - 49 =$ ______

Score

Space

Draw:

1 front view	2 side view	3 top view

4 How many cubes are needed for this model? ______

Length

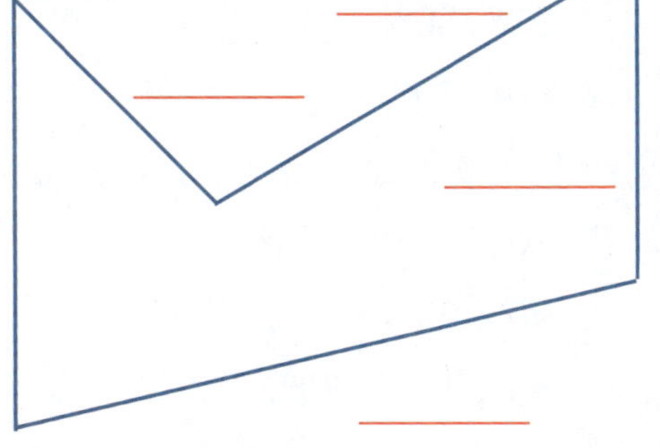

1 Measure each side in mm.

2 What is the perimeter? ______

Problem of the week

Using the numbers 1-9 find the value of each letter.

$A \times B = C$

$A + C = D$

$B \times E = FE$

$A \times G = FH$

$FG - D = I$

A	
B	
C	
D	
E	
F	
G	
H	
I	

Unit 25

A

1. \$2.20 ÷ 2 = ______
2. \$6.30 ÷ 3 = ______
3. \$10.50 ÷ 5 = ______
4. \$8.80 ÷ 4 = ______
5. \$18.60 ÷ 6 = ______
6. \$9.36 ÷ 3 = ______
7. \$12.84 ÷ 4 = ______
8. 6·8 ÷ 2 = ______
9. 9·6 ÷ 3 = ______
10. 5·5 ÷ 5 = ______
11. 4·8 ÷ 4 = ______
12. 7·2 ÷ 6 = ______
13. 9·6 ÷ 8 = ______
14. 12·6 ÷ 9 = ______
15. 10·5 ÷ 7 = ______
16. 14·4 ÷ 8 = ______
17. 15·6 ÷ 6 = ______
18. 29·7 ÷ 9 = ______
19. 34·3 ÷ 7 = ______
20. 41·5 ÷ 5 = ______

Score

B

1. $\frac{3}{4}$ + ______ = 1
2. $\frac{1}{3}$ + ______ = 1
3. $\frac{2}{5}$ + ______ = 1
4. $\frac{1}{6}$ + ______ = 1
5. $\frac{7}{10}$ + ______ = 1
6. $\frac{4}{5}$ + ______ = 1
7. $\frac{4}{10}$ + ______ = 1
8. $\frac{1}{8}$ + ______ = 1

Circle the larger.

9. $\frac{1}{2}$ $\frac{1}{4}$
10. $\frac{2}{3}$ $\frac{1}{2}$
11. $\frac{3}{10}$ $\frac{3}{5}$
12. $\frac{1}{4}$ $\frac{3}{8}$
13. $\frac{1}{2}$ $\frac{4}{10}$
14. $\frac{1}{3}$ $\frac{1}{4}$
15. $\frac{3}{4}$ $\frac{7}{8}$
16. $\frac{2}{5}$ $\frac{1}{2}$

Don't be tricked by the denominators!

Score

C

Name the type of angle.

1. 10° ______
2. 160° ______
3. 90° ______
4. 220° ______
5. 115° ______
6. 85° ______
7. 170° ______
8. 180° ______
9. 315° ______
10. 360° ______

Write the third angle of the triangle.

11. 60°, 70°, ______
12. 50°, 20°, ______
13. 90°, 80°, ______
14. 40°, 70°, ______

Write the fourth angle of the quadrilateral.

15. 100°, 120°, 50°, ______
16. 90°, 90°, 120°, ______
17. 50°, 130°, 80°, ______
18. 80°, 120°, 110°, ______

Score

Strategy

Jump strategy for + and –

eg 78 + 56 = 78 + 50 + 6 = 134	159 – 63 = 159 – 60 – 3 = 96

1 96 + 25 = ______
2 83 + 77 = ______
3 54 + 88 = ______
4 136 + 47 = ______
5 215 + 82 = ______

6 91 – 46 = ______
7 53 – 17 = ______
8 85 – 38 = ______
9 127 – 49 = ______
10 306 – 67 = ______

11 241 – 58 = ______
12 365 + 37 = ______
13 103 – 58 = ______
14 619 – 74 = ______
15 547 + 37 = ______

Score

Angles

Measure these angles.

1 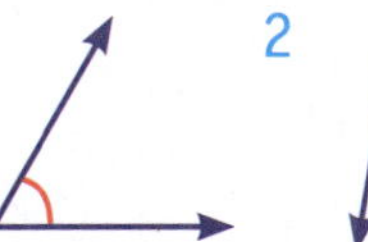 ______

2 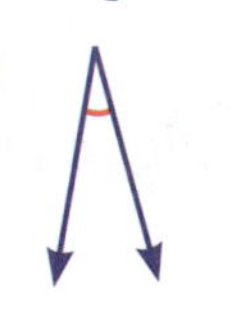 ______

3  ______

4  ______

Draw these angles.

5 125°

6 30°

7 82°

Calculator

Colour the calculations which give the answer.

1 51·4

A $\frac{3}{5}$ of 81	B 113·2 – 62·6
C 8·61 × 4	D 24·45 + 26·95

2 72·9

A 18·3 × 4	B 214·5 – 141·6
C $\frac{3}{4}$ of 97·2	D 16·8 + 55·1

Problem of the week

Lion, Chicken, Duck and Monkey are on an island with one two-man canoe. Chicken and Duck cannot be left alone with Lion. (He will eat them!) How can all the animals get to the mainland?

Unit 26

A

1 $3 \div 3 =$ ______
2 $5 \times 9 =$ ______
3 $16 \div 4 =$ ______
4 $7 \times 7 =$ ______
5 $35 \div 7 =$ ______
6 $6 \times 8 =$ ______
7 $81 \div 9 =$ ______
8 $36 \div 6 =$ ______
9 $28 \div 4 =$ ______
10 $8 \times 3 =$ ______
11 $72 \div 8 =$ ______
12 $9 \times 1 =$ ______
13 $100 \div 10 =$ ______
14 $6 \times 7 =$ ______
15 $8 \times 4 =$ ______
16 $56 \div 7 =$ ______
17 $7 \times 0 =$ ______
18 $9 \times 7 =$ ______
19 $54 \div 6 =$ ______
20 $18 \div 9 =$ ______

Score

B Draw lines to match.

1	$\frac{1}{2}$	a	$\frac{1}{4}$
2	$\frac{1}{10}$	b	0·5
3	0·25	c	9%
4	0·09	d	0·10
5	0·65	e	75%
6	$\frac{3}{4}$	f	0·20
7	20%	g	100%
8	1	h	65%

Write using Roman numerals.

9 14 ______
10 38 ______
11 65 ______
12 101 ______

Write using Hindu–Arabic numerals.

13 LXXII ______
14 XCVI ______
15 CCC ______

Score

C Would you use cm², m², ha?

1 area of book cover ______
2 area of classroom floor ______
3 area of garden ______
4 area of large paddock ______
5 area of national park ______
6 area of hand print ______
7 area of country town ______
8 area of tennis court ______
9 area of cereal box ______
10 area of TV screen ______

Perimeter of these rectangles.

11 length 7 cm width 4 cm ______
12 length 3 m width 2 m ______
13 length 12 mm width 7 mm ______
14 length 10 cm width 8 cm ______
15 length 15 m width 4 m ______
16 length 22 m width 9 m ______
17 length 16 mm width 7 mm ______
18 length 14 cm width 9 cm ______
19 length 7·2 m width 3·6 m ______
20 length $15\frac{1}{2}$ cm width $8\frac{1}{2}$ cm ______

Score

Targeting Mental Maths Year 5 • 978-1-922887-27-6

Strategy

To multiply by teens

× by 10 + × by 'teen' number.
eg 34 × 13 = 34 × 10 + 34 × 3
(= 340 + 102)
= 442

1 23 × 13 = 230 + 39
= ______

2 8 × 15 = ______
= ______

3 11 × 16 = ______
= ______

4 12 × 18 = ______
= ______

5 21 × 17 = ______
= ______

6 32 × 14 = ______
= ______

7 54 × 12 = ______
= ______

8 26 × 15 = ______
= ______

Score

Coordinates

1 Join (4,1) to (3,2) to (9, 2) to (8, 1) to (4, 1)

2 Join (5,2) to (5,3) to (0, 3) to (3, 5) to (1, 5) to (4,7) to (2, 7) to (6, 10) to (10, 7) to (8, 7) to (11, 5) to (9, 5) to (12, 3) to (7, 3) to (7, 2)

3 Colour the shape.

4 What have you drawn?

______

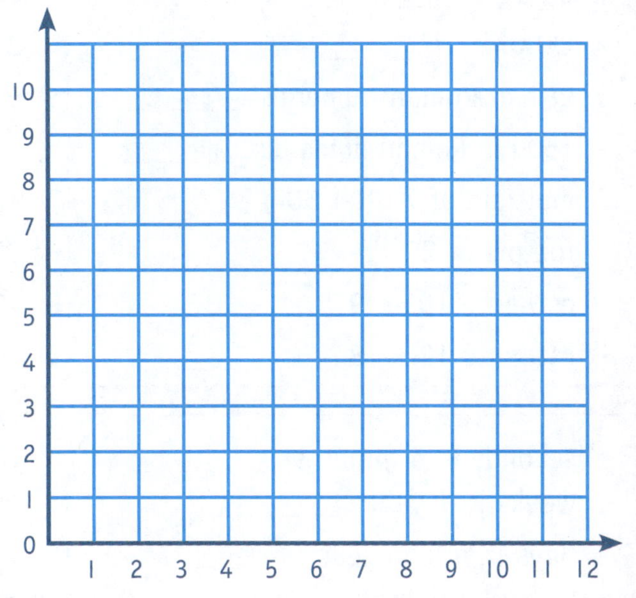

Problem of the week

Bill's Ice-cream Parlour sells 8 different flavours of ice-cream.

Jonas wants a double scoop.

How many different combinations could he choose? ______

Hint: Chocolate and Vanilla is the same as Vanilla and Chocolate; and don't forget doubles of the same flavour.

Unit 27 Revision

A

1 $9 \times 6 =$ ______
2 $8 + 14 + 12 =$ ______
3 $16 - 9 - 4 =$ ______
4 $42 \div 6 =$ ______
5 $7 \times 4 + 6 =$ ______
6 $33{\cdot}6 \div 8 =$ ______
7 $8 \times 7 =$ ______
8 $9 \times 7 =$ ______
9 $3 \times 8 + 15 =$ ______
10 $19 + 8 + 21 =$ ______
11 $6 \times 1 =$ ______
12 $11{\cdot}2 + 5{\cdot}9 =$ ______
13 $21 - 14 - 6 =$ ______
14 $23 + 16 + 7 =$ ______
15 $6{\cdot}4 + 4{\cdot}7 =$ ______
16 $8 \times 9 + 7 =$ ______
17 $10 \times 0 =$ ______
18 $33 - 5 - 8 =$ ______
19 $7.20 ÷ 4 = ______
20 $6 \times 8 =$ ______

Score

B

1 value of 2 in 52 703 ______
2 value of 2 in 214 995 ______
3 DCCIV = ______
4 LXIX = ______
5 281 in Roman numerals ______
6 1500 in Roman numerals ______
7 multiple of 7 after 50 ______
8 factors of 20 ______
9 estimate 105 ÷ 9 ______
10 estimate 12 × 18 ______
11 $0.08 × 10 = ______
12 $3.45 × 10 = ______
13 64c × 20 = ______
14 34c × 30 = ______
15 51 ÷ 7 = ______
16 $\frac{5}{8}$ + ______ = 1
17 1, 4, 9, 16, ______
18 25% of 1 year ______
19 10% of $2 ______
20 40% of $5 ______

Score

C

1 seconds in 4 minutes ______
2 weeks in 1 year ______
3 days in a leap year ______
4 months in a century ______
5 average of 7, 5, 3, 9, 11 ______
6 $3.15 + ______ = $10
7 total 11, 6 and 14 ______
8 sum of 23 and 16 ______
9 next prime number after 32 ______
10 How many 5c in 70c? ______
11 $7.25 + ______ = $10
12 $14.70 + ______ = $20
13 product of 7 and 6 ______
14 71 less 27 ______
15 square 9 ______
16 cube 3 ______
17 type of angle: 29° ______
18 perimeter of 7 m square ______
19 $12.65 + ______ = $20
20 product of 3·6 and 2 ______

Score

Strategy

Use strategies you have learnt.

1 76 × 10 = ______
2 507 × 10 = ______
3 18 × 20 = ______
4 5 × 6 + 8 × 6 = ______
5 83 − 44 = ______
6 91 − 37 = ______
7 27 + 16 + 13 = ______
8 93 ÷ 10 = ______
9 821 ÷ 100 = ______
10 78 + 25 = ______
11 104 − 49 = ______
12 223 + 86 = ______
13 17 × 14 = ______
14 52 × 12 = ______
15 14 × 7 + 6 × 7 = ______

Score

Space

Draw me. I have 5 faces and 5 corners. 4 of my faces are triangles.

Number

Name the fraction coloured.

1 ______ 2 ______ 3 ______ 4 ______

Space

1 This model is built using centimetre cubes. How many cubes are used in this model?

______

2 Draw.

top view

front view

Calculator

1 Find five square numbers between 2000 and 3000.

______

2 Find three multiples of 3, 4 and 5 less than 200.

______

3 What number when squared equals 729?

______

Unit 28

A

1. 2 × 0 = ______
2. 4 × 9 = ______
3. 5 × 7 = ______
4. 9 × 6 = ______
5. 1 × 8 = ______
6. 7 × 8 = ______
7. 4 × 4 = ______
8. 3 × 5 = ______
9. 9 × 7 = ______
10. 4 × 8 = ______
11. 5 × 5 = ______
12. 7 × 6 = ______
13. 6 × 6 = ______
14. 4 × 6 = ______
15. 4 × 7 = ______
16. 5 × 8 = ______
17. 8 × 10 = ______
18. 3 × 7 = ______
19. 9 × 9 = ______
20. 6 × 7 = ______

Score

B

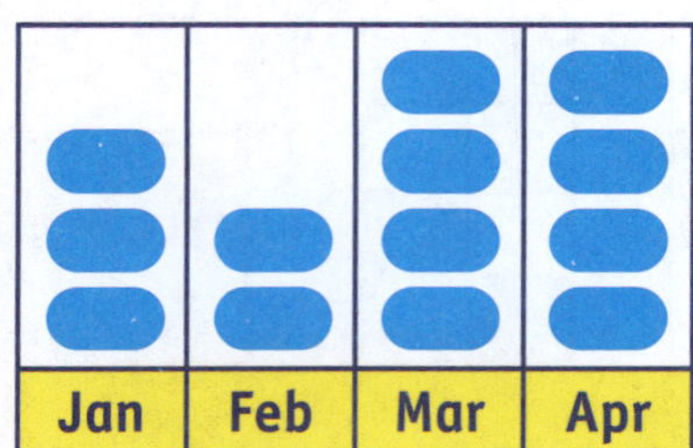

Scale = 6

1. How many in April? ______
2. Which month has twice as many as another month? ______
3. What question might have been asked to collect this data? ______
4. What would be a good title for this graph? ______

Score

C Write the value of the underlined digit in words.

1. 5 00$\underline{6}$ 703 ______
2. 4$\underline{5}$6 313 ______
3. 21·$\underline{3}$6 ______
4. 2$\underline{7}$·73 ______
5. 136·1$\underline{4}$ ______
6. 563·6$\underline{5}$3 ______
7. 104·7$\underline{0}$4 ______
8. 39$\underline{0}$·365 ______
9. 168·72$\underline{5}$ ______
10. 139·$\underline{7}$68 ______

Score

Unit 28

Strategy

To multiply by teens

× by 10 then by other number.
eg 35 × 16 (= 35 × 10 + 35 × 6)
= 350 + 210
= 560

1 23 × 14 = 200 + 92
= ______

2 31 × 16 = ______
= ______

3 27 × 15 = ______
= ______

4 44 × 13 = ______
= ______

5 62 × 17 = ______
= ______

6 75 × 12 = ______
= ______

7 53 × 18 = ______
= ______

8 39 × 13 = ______
= ______

9 62 × 19 = ______
= ______

Score

Number

Colour to show:

1 $3 - \frac{3}{4} =$ ______

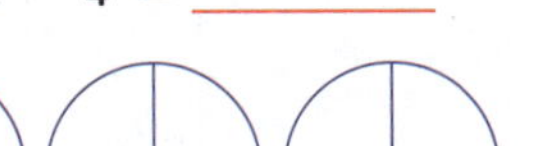

2 $2 - \frac{3}{8} =$ ______

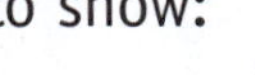

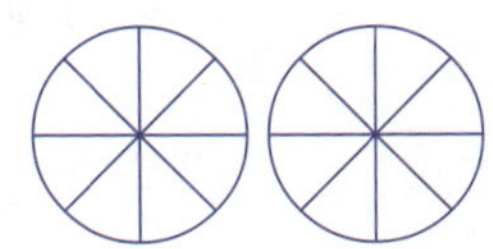

3 $4 - \frac{7}{10} =$ ______

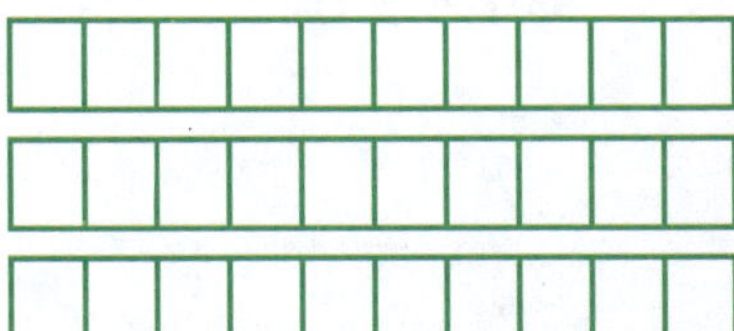

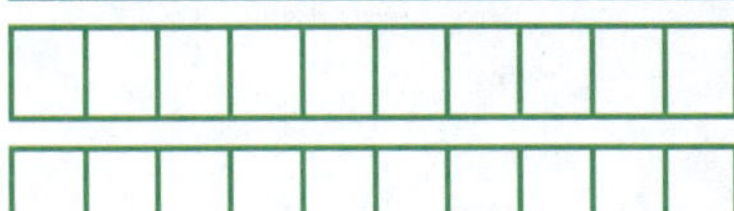

4 $2 - \frac{2}{5} =$ ______

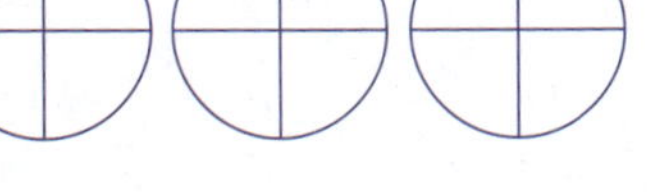

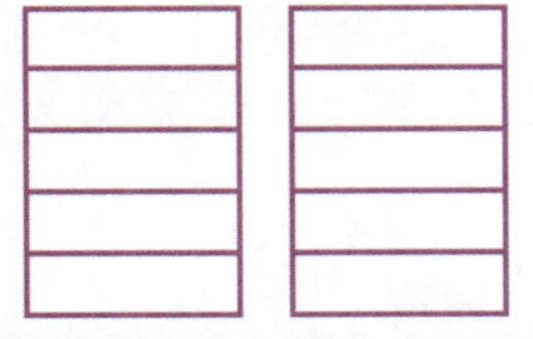

5 $3 - 1\frac{2}{3} =$ ______

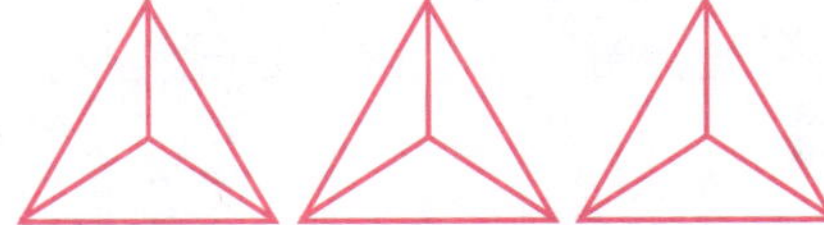

Space

Name a shape with:

1 4 sides. ______

2 5 sides. ______

3 6 sides. ______

4 7 sides. ______

5 8 sides. ______

6 9 sides. ______

7 10 sides. ______

Problem of the week

Contact lenses were invented 13 years before the start of the 20th century.

Eyeglasses were invented 602 years earlier than that.

In what year were eyeglasses invented?

______

Unit 29

A

1. 45 ÷ 5 = ______
2. 36 ÷ 6 = ______
3. 72 ÷ 8 = ______
4. 9 ÷ 1 = ______
5. 24 ÷ 4 = ______
6. 64 ÷ 8 = ______
7. 18 ÷ 2 = ______
8. 10 ÷ 2 = ______
9. 25 ÷ 5 = ______
10. 12 ÷ 3 = ______
11. 49 ÷ 7 = ______
12. 80 ÷ 10 = ______
13. 81 ÷ 9 = ______
14. 48 ÷ 6 = ______
15. 18 ÷ 3 = ______
16. 56 ÷ 7 = ______
17. 54 ÷ 6 = ______
18. 27 ÷ 9 = ______
19. 20 ÷ 4 = ______
20. 42 ÷ 6 = ______

Score

B

1. $8\overline{)792}$
2. $4\overline{)560}$
3. $6\overline{)774}$
4. $9\overline{)891}$
5. $5\overline{)645}$
6. $3\overline{)852}$
7. $6\overline{)738}$
8. $8\overline{)928}$
9. $7\overline{)567}$
10. $5\overline{)1365}$

Score

C Write the missing word.

1. odd × odd = ______
2. ______ × even = even
3. ______ × odd = even
4. odd × odd × odd = ______
5. odd × ______ × even =even
6. even × odd × even = ______
7. odd + odd = ______
8. even + odd = ______
9. odd + ______ + odd = even
10. even + even + ______ = odd
11. next multiple of 8 after 96 ______
12. factors of 4. ______, ______ and ______
13. remainder of 39 ÷ 4 ______
14. remainder of 86 ÷ 12 ______
15. degrees in a right angle ______
16. next multiple of 6 after 66 ______
17. A horse weighs about 500 ______.
18. A hallway is about 15 ______ long.
19. A cup holds about 200 ______.
20. A pin head is about 2 ______ wide.

Score

Targeting Mental Maths Year 5 • 978-1-922887-27-6

Unit 29

Strategy
Estimation – round to easy numbers

eg 68 + 57 → 70 + 60 → estimate 130
89 × 17 → 100 × 20 → estimate 2000

Estimate answers.

1 84 – 29 ________
2 18 × 9 ________
3 139 – 48 ________
4 215 ÷ 8 ________
5 37 × 6 ________
6 246 – 138 ________
7 93 + 126 ________
8 297 ÷ 53 ________
9 8 × 9 × 13 = ________
10 64 + 43 + 92 = ________
11 308 – 149 + 67 = ________
12 3213 × 19 = ________

Score

Space

Name the type of angle. Measure each angle.

1
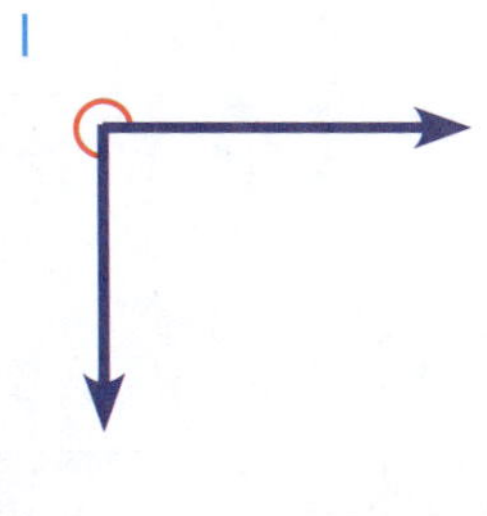

2
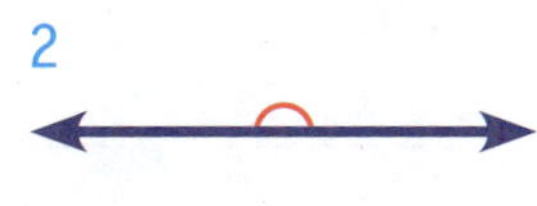

3
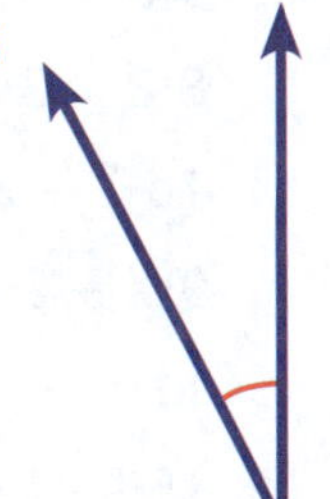

4
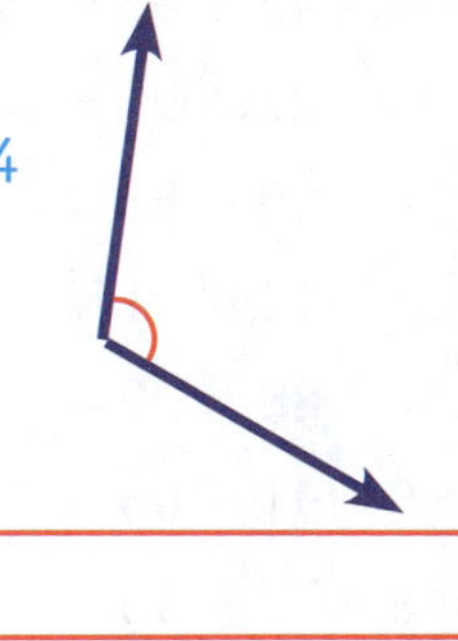

Time

Write the 12-hour time.

1 15:23 ________
2 07:02 ________
3 20:51 ________
4 11:19 ________
5 16:48 ________

Problem of the week

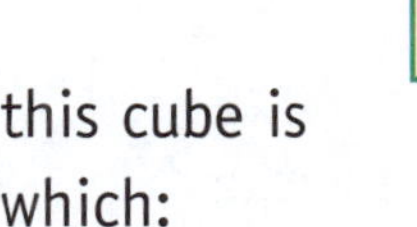

When this cube is made which:

1 shape is opposite the triangle? ________
2 shapes are next to the triangle? ________
3 shape is the arrow pointing to? ________
4 shape is not next to the dot? ________

Unit 30

A

1 4 × 8 − 5 = ________
2 2 × 7 + 11 = ________
3 4 × 6 − 9 = ________
4 7 × 3 + 12 = ________
5 9 × 8 − 5 = ________
6 6 × 9 − 8 = ________
7 8 × 7 + 6 = ________
8 7 × 4 − 13 = ________
9 9 × 5 + 11 = ________
10 8 × 8 − 15 = ________
11 7 × 9 − 8 = ________
12 6 × 0 + 5 = ________
13 6 × 10 − 7 = ________
14 8 × 5 + 15 = ________
15 8 × 6 − 13 = ________
16 7 × 7 + 8 = ________
17 3 × 8 + 15 = ________
18 10 × 10 − 17 = ________
19 6 × 8 − 21 = ________
20 9 × 4 + 14 = ________

Score

B

Estimate only.

1 79 + 16 = ________
2 43 + 85 = ________
3 27 + 99 = ________
4 66 + 53 = ________
5 144 + 162 = ________
6 88 − 29 = ________
7 91 − 65 = ________
8 73 − 38 = ________

Estimate in whole numbers. Then write the answer.

9 1·7 + 2·2 = ________ ________
10 3·5 + 5·1 = ________ ________
11 9·8 + 6·4 = ________ ________
12 8·2 + 3·7 = ________ ________
13 16·3 + 21·8 = ________ ________
14 9·7 − 2·5 = ________ ________
15 11·3 − 3·1 = ________ ________
16 15·9 − 8·6 = ________ ________

Score

C

Circle the higher temperature.

1 16°C 24°C
2 −3°C 1°C
3 5°C −7°C
4 0°C 10°C
5 2°C −4°C

What would you measure with:

6 a tape measure? ________
7 kitchen scales? ________
8 a thermometer? ________
9 an odometer? ________
10 a stopwatch? ________

Write in words.

11 12 000 000 ________
12 3 000 000 ________
13 100 000 ________
14 15 000 005 ________
15 1 000 100 ________

Write using numerals.

16 two million and five thousand ________
17 eight million and ten ________
18 five hundred thousand ________
19 fourteen million ________
20 one million one hundred ________

Score

Targeting Mental Maths Year 5 • 978-1-922887-27-6

Strategy

Look for patterns

eg 6 + 7 = 13
60 + 70 = 130
600 + 700 = 1300

1 9 + 8 = ______
90 + 80 = ______
900 + 800 = ______

2 5 + 7 = ______
50 + 70 = ______
500 + 700 = ______

3 13 + 8 = ______
130 + 80 = ______
1300 + 800 = ______

4 15 + 9 = ______
150 + 90 = ______
1500 + 900 = ______

5 7 + 19 = ______
70 + 190 = ______
700 + 1900 = ______

6 6 + 8 = ______
600 + 800 = ______
6000 + 8000 = ______

Score

Measurement

Find the perimeter of each of these squares.

1 Area = 9 cm^2

perimeter = ______

2 Area = 16 cm^2

perimeter = ______

3 Area = 25 cm^2

perimeter = ______

4 Area = 36 cm^2

perimeter = ______

Position

Write the direction of:

1 A from E. ______
2 E from B. ______
3 H from F. ______
4 E from A. ______
5 C from G. ______

Problem of the week

This is $\frac{1}{4}$ of a shape.
Draw the whole shape.

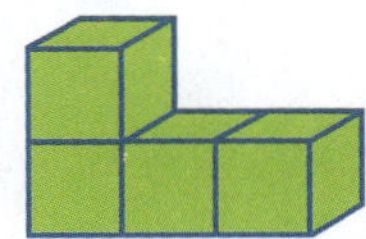

Unit 31

A

1 13 + 9 + 7 = ______
2 11 + 14 + 6 = ______
3 15 + 5 + 8 = ______
4 19 + 7 + 1 = ______
5 18 + 18 + 2 = ______
6 7 + 17 + 3 = ______
7 12 + 6 + 8 = ______
8 14 + 7 + 16 = ______
9 10 + 13 + 10 = ______
10 9 + 12 + 18 = ______
11 11 + 19 + 4 = ______
12 7 + 12 + 13 = ______
13 6 + 9 + 24 = ______
14 5 + 3 + 27 = ______
15 35 + 5 + 8 = ______
16 13 + 9 + 21 = ______
17 4 + 36 + 11 = ______
18 8 + 9 + 22 = ______
19 26 + 13 + 4 = ______
20 51 + 17 + 49 = ______

Score

B

1 $0.35 × 10 = ______
2 $0.07 × 10 = ______
3 $0.48 × 10 = ______
4 $0.15 × 10 = ______
5 $0.92 × 10 = ______
6 $3.40 × 10 = ______
7 $1.86 × 10 = ______
8 $5.29 × 10 = ______
9 $8.03 × 10 = ______
10 $11.67 × 10 = ______
11 49c × 20 = ______
12 56c × 20 = ______
13 15c × 20 = ______
14 25c × 20 = ______
15 35c × 30 = ______
16 83c × 30 = ______
17 60c × 40 = ______
18 75c × 40 = ______
19 $1.10 × 20 = ______
20 $2.30 × 30 = ______

Score

C Complete these equivalent fractions.

1 $\frac{\square}{10} = 0{\cdot}75$

2 $\frac{4}{8} = \frac{1}{\square}$

3 $\frac{1}{2} = \frac{\square}{6}$

4 $\frac{7}{10} = 0{\cdot}$______

5 $\frac{3}{5} = \frac{\square}{10}$

Write a chance word that means:

6 is sure to happen. ______
7 maybe. ______
8 more than likely. ______
9 one chance in two. ______
10 highly improbable. ______
11 will never happen. ______
12 just a slight chance. ______
13 without a doubt. ______
14 is almost sure. ______
15 fifty-fifty. ______

Score

Targeting Mental Maths Year 5 • 978-1-922887-27-6

Strategy

Split strategy for addition

eg 67 + 37 = 60 + 30 + 7 + 7
= 104

1 39 + 46 = ______
2 58 + 33 = ______
3 62 + 71 = ______
4 84 + 19 = ______
5 73 + 28 = ______
6 47 + 73 = ______
7 61 + 58 = ______
8 90 + 27 = ______
9 18 + 74 = ______
10 35 + 59 = ______
11 26 + 48 = ______
12 17 + 54 = ______
13 89 + 17 = ______
14 55 + 36 = ______
15 68 + 25 = ______

Score

Number

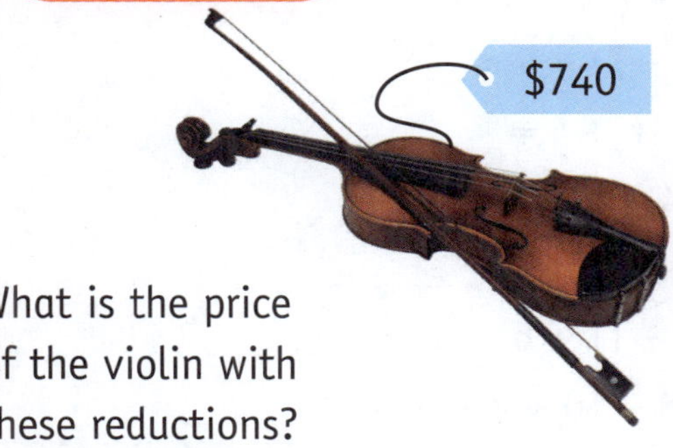

What is the price of the violin with these reductions?

1 50% ______
2 25% ______
3 10% ______
4 40% ______
5 75% ______
6 12% ______

Money

What is the best buy?

1

$12.60

a 50% off
b $6.50 off

______

2 $80

a reduced by $25
b 25% off

______

3 $16.80

a reduced by $\frac{1}{4}$
b $4.25 taken off

______

Problem of the week

In the triangle the angle at A is 40° less than the angle at B and 20° less than the angle at C.

What is the size of each angle?

∠A = ______ ∠B = ______ ∠C = ______

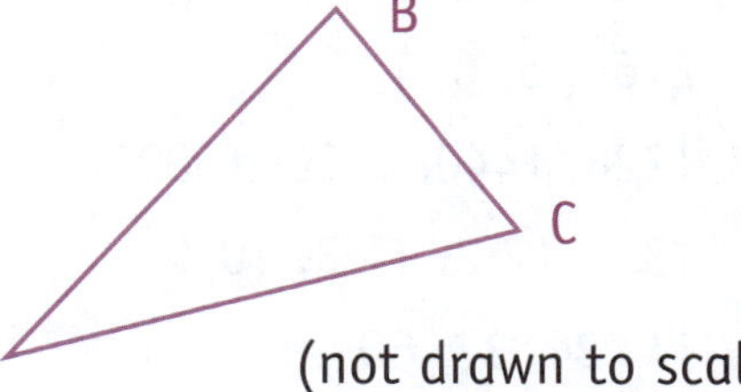

(not drawn to scale)

Unit 32

A

1 16 + 5 = ______
2 36 − 3 = ______
3 48 + 7 = ______
4 62 + 8 = ______
5 51 − 4 = ______
6 93 − 7 = ______
7 38 + 8 = ______
8 26 + 9 = ______
9 37 − 9 = ______
10 58 + 9 = ______
11 94 − 7 = ______
12 87 − 8 = ______
13 65 + 5 = ______
14 35 − 8 = ______
15 89 − 4 = ______
16 76 + 7 = ______
17 79 − 6 = ______
18 66 + 5 = ______
19 89 − 7 = ______
20 52 + 9 = ______

Score

B

1 7 + 5 × 2 = ______
2 8 ÷ 4 + 7 = ______
3 15 − 12 ÷ 3 = ______
4 6 × 5 + 2 = ______
5 27 ÷ 3 − 7 = ______
6 4 × 2 + 3 × 3 = ______
7 18 ÷ 9 + 6 × 1 = ______
8 (3 + 4) × 5 = ______
9 8 − 28 ÷ 4 = ______
10 (15 − 12) × 9 = ______
11 7 + 8 ÷ 4 − 5 = ______
12 8 − 4 + 6 × 3 = ______
13 (3 + 2) × (12 − 6) = ______
14 9 × (12 − 7) − 4 = ______
15 3 × 7 + 16 ÷ 4 = ______
16 14 × (14 − 4) = ______
17 12 + 8 − 18 ÷ 6 = ______
18 (12 + 9 − 11 + 6) × 2 = ______
19 10 ÷ (18 − 10 + 2 − 5) = ______
20 36 ÷ 9 + 4 × 3 = ______

Score

C Match each pattern with its rule.

1 3, 9, 27, 81
2 0·1, 0·7, 1·3, 1·9
3 $2\frac{2}{3}$, $2\frac{1}{3}$, 2, $1\frac{2}{3}$
4 6, 3, 1·5, 0·75
5 $2\frac{1}{2}$, $3\frac{1}{4}$, 4, $4\frac{3}{4}$
6 1·8, 3·5, 5·2, 6·9
7 216, 36, 6, 1
8 1728, 1440, 1200, 1000
9 13·1, 12·2, 11·3, 10·4
10 11, 24, 37, 50

a $-\frac{1}{3}$
b × 0·5
c + 1·7
d × 3
e ÷ 6
f + 0·6
g − 0·9
h $+\frac{3}{4}$
i + 13
j ÷ 1·2

Perimeter and area of these rectangles.

11 length 6 cm, width 9 cm
P = ______ A = ______
12 length 9 m, width 7 m
P = ______ A = ______
13 length 11 mm, width 9 mm
P = ______ A = ______
14 length 20 km, width 8 km
P = ______ A = ______
15 length 17 cm, width 7 cm
P = ______ A = ______

Score

Unit 32

Strategy
Near doubles

44 + 47 = 44 + 44 + 3
= 91

35 + 33 = 35 + 35 − 2
= 68

1 15 + 17 = ________
2 14 + 12 = ________
3 23 + 25 = ________
4 32 + 31 = ________
5 43 + 45 = ________
6 27 + 25 = ________
7 31 + 29 = ________
8 43 + 46 = ________
9 19 + 18 = ________
10 56 + 55 = ________
11 130 + 129 = ________
12 81 + 82 = ________
13 45 + 47 = ________
14 62 + 64 = ________
15 16 + 18 = ________

Score

Number

Complete the decimal number line.

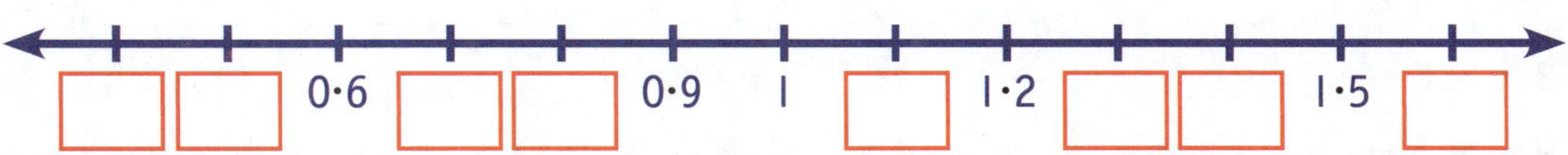

Data

Grandad was comparing the price of new suits over the last ten years of his working life. Starting from 1992 these were the prices:
$150, $175, $225, $225, $250,
$290, $300, $250, $315, $350.

1 Draw the line graph to show this information.

2 What do you think happened in 1999?

____

Problem of the week

When you multiply one outside number by one inside number the digits of their product total 9.

1 What are the numbers? ________

2 Which outside number and which inside number give a product closest to 1200? ________

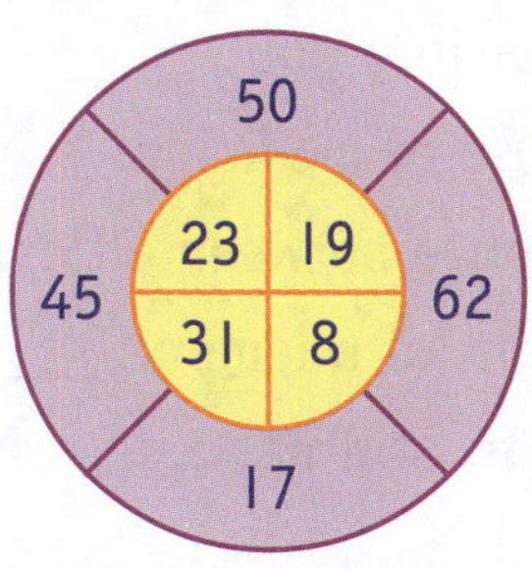

Unit 33

A

1. 1·3 + 1·2 = ________
2. 2·7 + 1·1 = ________
3. 3·8 + 2·9 = ________
4. 4·6 + 5·3 = ________
5. 8·4 + 7·7 = ________
6. 6·9 + 5·8 = ________
7. 3·5 + 4·9 = ________
8. 8·6 – 6·3 = ________
9. 9·8 – 4·5 = ________
10. 10·7 – 2·7 = ________
11. 8·4 – 3·6 = ________
12. 3·5 – 1·8 = ________
13. 7·1 – 2·9 = ________
14. 6·2 – 5·5 = ________
15. 4·3 – 0·8 = ________
16. 10·4 – 6·4 = ________
17. 12·1 – 7·2 = ________
18. 8·3 – 7·5 = ________
19. 3·9 + 7 = ________
20. 8 – 2·4 = ________

Score

B

1. \$4 × 6 = ________
2. \$9 × 9 = ________
3. 55c × 3 = ________
4. 85c × 4 = ________
5. 31c × 6 = ________
6. 95c × 8 = ________
7. \$1.20 × 7 = ________
8. \$5.40 × 5 = ________
9. \$7.60 × 3 = ________
10. \$8.15 × 9 = ________

Write the value of the star.

11. ★ + 3 = 4 × 2 ________
12. 9 – ★ = 36 ÷ 6 ________
13. 50 + ★ = 9 × 6 ________
14. ★ ÷ 7 = 42 ÷ 6 ________
15. 9 × 3 = ★ – 3 ________
16. 48 ÷ 8 = 15 – ★ ________
17. 13 + ★ = 9 × 7 ________
18. 100 – ★ = 9^2 ________

Score

C

Write the measure used for:

1. milk in a cup. ________
2. weight of a tiger. ________
3. length of a book. ________
4. length of a lesson. ________
5. length of a pin. ________
6. water in a pool. ________
7. a dose of medicine. ________
8. distance between cities. ________
9. weight of a lemon. ________
10. length of a park. ________

Estimate.

11. length of your classroom ________
12. height of your teacher ________
13. weight of a new baby kitten ________
14. time from sunrise to midday ________
15. weight of a bucket of oranges ________
16. time to say the alphabet ________
17. milk in a glass ________
18. width of a pencil ________
19. water in a bucket ________
20. length of a toothpaste tube ________

Score

Targeting Mental Maths Year 5 • 978-1-922887-27-6

Strategy

÷ by 4
÷ by 8

To divide by 4, halve and halve again.	To divide by 8, halve, halve and halve again.
eg 76 ÷ 4 (= 76 ÷ 2 ÷ 2) = 19	eg 184 ÷ 8 (= 184 ÷ 2 ÷ 2 ÷ 2) = 23

1 52 ÷ 4 = ______
2 96 ÷ 4 = ______
3 132 ÷ 4 = ______
4 216 ÷ 4 = ______
5 96 ÷ 8 = ______
6 120 ÷ 8 = ______
7 168 ÷ 8 = ______
8 280 ÷ 8 = ______
9 176 ÷ 4 = ______
10 152 ÷ 8 = ______
11 1640 ÷ 8 = ______
12 1020 ÷ 4 = ______

Score

Measurement

Find the area of each shape.

1

3 m
2 m
2 m
8 m

A =

2

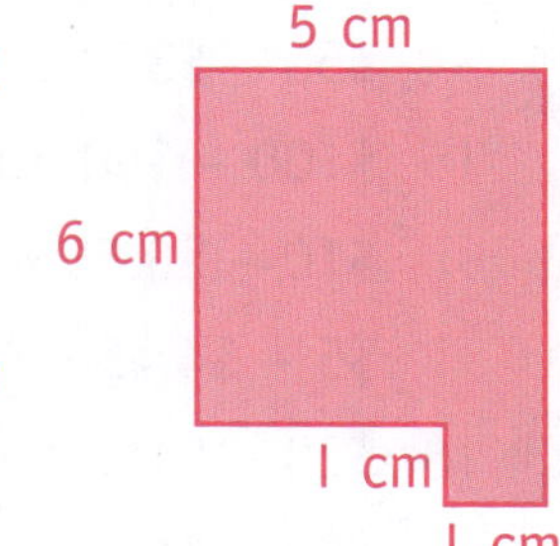

A =

3

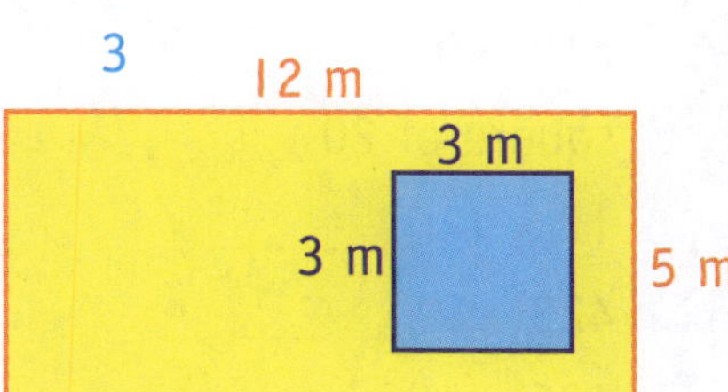

A =

Position

F	A	G
B	E	D
H	C	I

Which letter is:

1 North of E. ______
2 South of B. ______
3 East of E. ______
4 Northeast of E. ______
5 West of A. ______
6 Southeast of B. ______
7 Northwest of C. ______
8 Southwest of E. ______
9 West of C. ______
10 Between F and G. ______

Problem of the week

Chocolates are packed in layers. Each layer has 4 rows of 8 chocolates.
How many layers are needed to pack a box with 128 chocolates? ______

Unit 34

A

1 8 × 2 = ______

2 7 × 1 = ______

3 72 ÷ 9 = ______

4 4 × 4 = ______

5 7 × 8 = ______

6 10 × 10 = ______

7 21 ÷ 3 = ______

8 3 × 8 = ______

9 45 ÷ 5 = ______

10 9 ÷ 1 = ______

11 7 × 6 = ______

12 18 ÷ 2 = ______

13 63 ÷ 7 = ______

14 40 ÷ 5 = ______

15 9 ÷ 3 = ______

16 6 × 9 = ______

17 4 × 6 = ______

18 81 ÷ 9 = ______

19 5 × 7 = ______

20 28 ÷ 4 = ______

Score

B

1 23 × 5 ______

2 94 + 58 ______

3 218 – 68 ______

4 315 ÷ 6 ______

5 square of 20 ______

6 17 × 11 ______

7 423 ÷ 9 ______

8 126 + 188 ______

9 540 – 139 ______

10 58 + 37 + 42 ______

11 $10 – $7.50 = ______

12 $20 – $11.30 = ______

13 $5 – $2.95 = ______

14 $50 – $35.60 = ______

15 $100 – $60.80 = ______

16 $10 – $0.55 = ______

17 $5 – $4.45 = ______

18 $20 – $18.85 = ______

19 $50 – $6.10 = ______

20 $10 – $9.55 = ______

Score

C

What am I?

1 I have 2 circular ends. ______

2 I have 5 faces, two of which are triangles. ______

3 I have 5 faces, 4 of which are triangles. ______

4 I have 1 flat surface and 1 curved surface. ______

5 I have 6 identical faces. ______

How many:

6 feet on 16 chickens? ______

7 cents in 7·5 dollars? ______

8 faces on 3 cubes? ______

9 years between leap years? ______

10 degrees in a straight angle? ______

11 minutes in $2\frac{1}{4}$ hours? ______

12 mL in $1\frac{3}{4}$ L? ______

13 seconds in $5\frac{1}{2}$ minutes? ______

14 cm in $19\frac{1}{2}$ m? ______

15 days in spring? ______

Score

Unit 34

Strategy
× by 25

To multiply by 25 – × 100 then ÷ 4
eg 16 × 25 = 1600 ÷ 4
= 400

1 12 × 25 = 1200 ÷ 4
= ______

2 20 × 25 = ______
= ______

3 24 × 25 = ______
= ______

4 44 × 25 = ______
= ______

5 36 × 25 = ______
= ______

6 28 × 25 = ______
= ______

7 14 × 25 = ______
= ______

8 21 × 25 = ______
= ______

9 23 × 25 = ______
= ______

Score

Space

1 Measure this angle.

______

2 What type of angle is it?

______

3 Draw an angle of 210°.

Position

Write the coordinates.

1 A ______

2 B ______

3 C ______

4 D ______

5 E ______

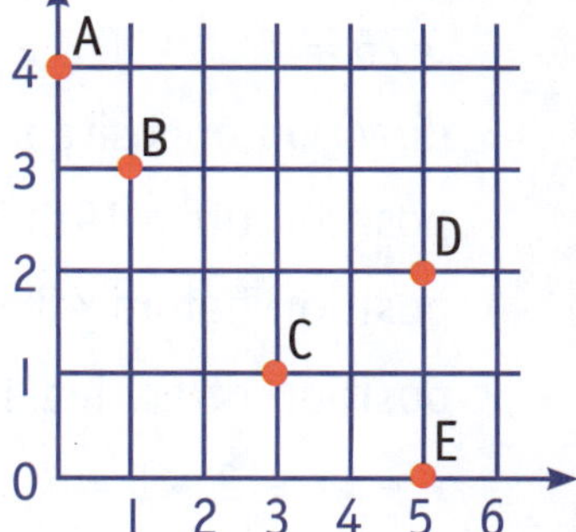

Space

Draw all the diagonals.

Problem of the week

I am a three-digit number. I am divisible by 3. My hundreds digit is is one third my tens digit. If you reverse me I am divisible by 6.

I am ______.

Unit 35 Revision

A

1 7 × 6 = ________
2 16 + 9 + 14 = ________
3 5 × 9 = ________
4 94 − 7 = ________
5 10 ÷ 10 = ________
6 12 + 14 + 8 = ________
7 4 × 3 + 13 = ________
8 4 × 9 = ________
9 7 × 7 + 9 = ________
10 79 − 6 = ________
11 9 × 8 = ________
12 1·7 + 2·5 = ________
13 56 ÷ 7 = ________
14 10·2 − 3·8 = ________
15 8·4 + 6·9 = ________
16 48 ÷ 6 = ________
17 62 + 8 = ________
18 6 × 6 − 17 = ________
19 20 ÷ 10 = ________
20 10 × 10 − 37 = ________

Score

B

1 remainder of 21 ÷ 6 ________
2 value of 3 in 732 164 ________
3 $\frac{1}{2} = \frac{}{8}$
4 $0{\cdot}75 = \frac{}{4}$
5 direction opposite south ________
6 position after 19th ________
7 position before 4th ________
8 position after tied 10th ________
9 7 + 12 ÷ 3 − 5 = ________
10 Complete: 0·3, 0·7, 1·1, 1·5, ________
11 Complete: $3\frac{5}{8}$, $3\frac{3}{8}$, $3\frac{1}{8}$, ________
12 direction opposite south-east ________
13 write 3 multiples of 8 ________ ________ ________
14 write 3 factors of 42 ________ ________ ________
15 change to dollars: 1053c ________
16 419 ÷ 10 = ________
17 9 × (15 − 8) = ________
18 9 + ★ = 3 × 4 ★ = ________

Score

C

1 26·45 + 17·6 ________
2 49 ÷ 6 = ________ r ________
3 degrees in a straight line ________
4 20 cent coins in $1 ________
5 $\frac{1}{2} + \frac{3}{4} =$ ________
6 I have 6 square faces. ________
7 $10 − $3.15 = ________
8 $50 − $21.40 = ________
9 seconds in $3\frac{1}{2}$ mins ________
10 degrees in a straight angle ________
11 7 pm in 24-hour time ________
12 $475 k = ________
13 7·3 cm = ________ mm
14 winter = ________ days
15 circle the higher one 1°C, −3°C, 2°C
16 chance word for maybe ________
17 half an hour before 13:15 hrs = ________
18 rectangle of length 13 cm, width 6 cm
perimeter = ________ area = ________
19 estimate the length of the board ________
20 cents in $2\frac{1}{4}$ dollars ________

Score

Targeting Mental Maths Year 5 • 978-1-922887-27-6

Strategy

Use strategies you have learnt.

1 16 + 18 = ______

2 76 + 75 = ______

3 17 + 6 = ______

170 + 60 = ______

1700 + 600 = ______

4 estimate 43 × 19 = ______

5 estimate 137 + 65 = ______

6 estimate 431 ÷ 6 = ______

7 96 + 88 = ______

8 152 + 39 = ______

9 84 × 15 = ______

10 148 ÷ 4 = ______

11 288 ÷ 8 = ______

12 16 × 25 = ______

13 24 × 25 = ______

Score

Space

Name the type of angle. Measure each angle.

1 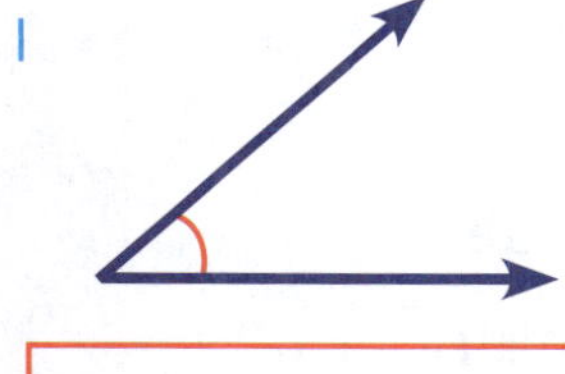

2 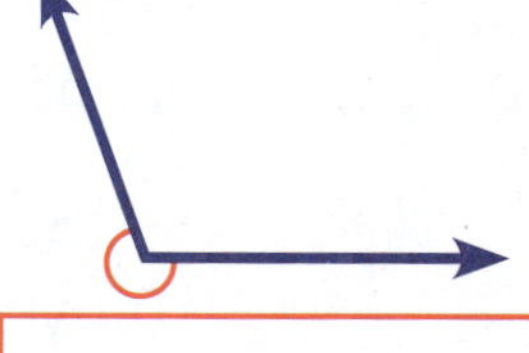

3 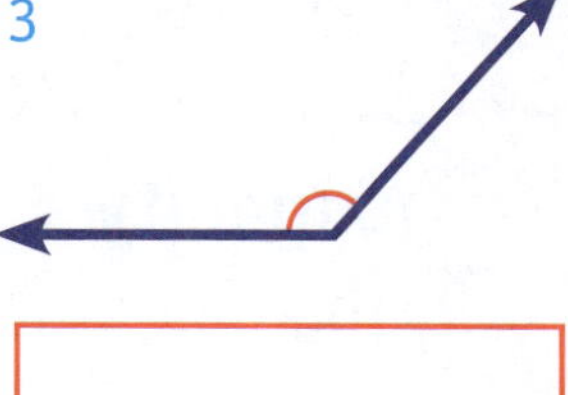

Measurement

1 Write the 12-hour time.

______

22:05 ______

2 These times are pm. Write the 24-hour time.

______ ______

Measurement

8 m

6 m

8 m

12 m

1 Perimeter = ______ 2 Area = ______

Number

What is the price with these reductions?

1 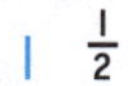 $\frac{1}{2}$ ______

2  $\frac{1}{10}$ ______

3 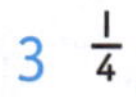 $\frac{1}{4}$ ______

4 $\frac{1}{5}$ ______

Measurement

Length

10 mm = 1 cm

100 cm = 1 m

1000 m = 1 km

Mass

1000 g = 1 kg

1000 kg = 1 t

Capacity

1000 mL = 1 L

1000 L = 1 kL

Area

100 mm^2 = 1 cm^2

10 000 cm^2 = 1 m^2

10 000 m^2 = 1 ha

100 ha = 1 km^2

Time

60 seconds = 1 minute

60 minutes = 1 hour

24 hours = 1 day

7 days = 1 week

14 days = 2 weeks = 1 fortnight

365 days = 1 year

366 days = 1 leap year

12 months = 1 year

10 years = 1 decade

100 years = 1 century

1000 years = 1 millennium

Days in months

30 days has September,
April, June and November.
All the rest have 31
Except February alone,
Which has 28 days clear
And 29 days each leap year.